Harrisowitz

Advances in Polymer Science

Fortschritte der Hochpolymeren-Forschung

Volume 15

Edited by

H.-J. Cantow, Freiburg i. Br. · G. Dall'Asta, Milano · J. D. Ferry, Madison · H. Fujita, Osaka · M. Gordon, Colchester · W. Kern, Mainz G. Natta, Milano · S. Okamura, Kyoto · C. G. Overberger, Ann Arbor W. Prins, Syracuse · G. V. Schulz, Mainz · W. P. Slichter, Murray Hill A. J. Staverman, Leiden · J. K. Stille, Iowa City

With 32 Figures

Springer-Verlag Berlin · Heidelberg · New York 1974

Editors

Prof. Dr. HANS-JOACHIM CANTOW, Institut für Makromolekulare Chemie der Universität, 7800 Freiburg i. Br., Stefan-Meier-Str. 31, BRD

Prof. Dr. GINO DALL'ASTA, Istituto di Chimica, Industriale del Politecnico, Piazza Leonardo da Vinci 32, Milano, Italia

Prof. Dr. JOHN D. FERRY, Department of Chemistry, The University of Wisconsin, Madison 6, Wisconsin 53706, USA

Prof. Dr. HIROSHI FUJITA, Osaka University, Department of Polymer Science, Toyonaka, Osaka, Japan

Prof. Dr. MANFRED GORDON, University of Essex, Department of Chemistry, Wivenhoe Park, Colchester C04 3SQ, England

Prof. Dr. WERNER KERN, Institut für Organische Chemie der Universität, 6500 Mainz, BRD

Prof. Dr. GIULIO NATTA, Istituto di Chimica Industriale del Politecnico, Milano, Italia

Prof. Dr. SEIZO OKAMURA, Department of Polymer Chemistry, Kyoto University, Kyoto, Japan

Prof. Dr. CHARLES G. OVERBERGER, The University of Michigan, Department of Chemistry, Ann Arbor, Michigan 48104, USA

Prof. Dr. WILLEM PRINS, Department of Chemistry, Syracuse University, Syracuse, N.Y. 13210, USA

Prof. Dr. GÜNTER VICTOR SCHULZ, Institut für Physikalische Chemie der Universität, 6500 Mainz, BRD

Dr. WILLIAM P. SLICHTER, Bell Telephone Laboratories Incorporated, Chemical Physics Research Department, Murray Hill, New Jersey 07971, USA

Prof. Dr. ALBERT JAN STAVERMAN, Chem. Laboratoria der Rijks-Universiteit, afd. Fysische Chemie I, Wassenaarseweg, Postbus 75, Leiden, Nederland

Prof. Dr. JOHN K. STILLE, University of Iowa, Department of Chemistry, Iowa City, USA

ISBN 3-540-06910-0 Springer-Verlag Berlin · Heidelberg · New York
ISBN 0-387-06910-0 Springer-Verlag New York · Heidelberg · Berlin

The use of general descrive names, trade marks, etc. in this publication, even if the former are not especially identified, is not to be taken as a sign that such names, as understood by the Trade Marks and Merchandise Marks Act, may accordingly be used freely by anyone.

This work is subject to copyright. All rights are reserved, whether the whole or part of the material is concerned, specifically those of translation, reprinting, re-use of illustrations, broadcasting, reproduction by photocopying, machine or similar means, and storage in data banks. Under § 54 of the German Copyright Law where copies are made for other than private use, a fee is payable to the publisher, the amount to the fee to be determined by agreement with the publisher. © by Springer-Verlag Berlin · Heidelberg 1974. Library of Congress Catalog Card Number 61-642. Printed in Germany. Typesetting and printing: Brühlsche Universitätsdruckerei, Gießen

Contents

Oligomerization of Ethylene with Soluble Transition-Metal Catalysts
   GISELA HENRICI-OLIVÉ and SALVADOR OLIVÉ 1

Stereochemistry of Propylene Polymerization
   ADOLFO ZAMBELLI and CAMILLO TOSI 31

Mercaptan-Containing Polymers
   CHIEN-DA S. LEE and WILLIAM H. DALY 61

Structures of Copolymers of High Olefins
   YURI V. KISSIN 91

Oligomerization of Ethylene with Soluble Transition-Metal Catalysts

G. HENRICI-OLIVÉ and S. OLIVÉ

Monsanto Research S. A., Zürich, Switzerland

Contents

1. Introduction 2
2. General Concepts 4
   2.1. The β-Hydrogen Abstraction 4
   2.2. Stability of the Metal toward Reduction 5
   2.3. Copolymerization of α-Olefins (Branching) 7
3. Catalyst Tailoring in Ethylene Oligomerization 8
   3.1. Ti-Al Catalysts 8
   3.2. Sulfur-Containing Additives 12
   3.3. Zr-Al Catalysts 14
   3.4. Al-Free Ti Catalysts 16
4. Kinetics and Mechanism 16
   4.1. Kinetic Equations and Conditions for their Application . . 16
   4.2. Evaluation of a Ti-Al System 19
   4.3. Comparison with other Systems 24
   4.4. Consequences of the Reaction Mechanism 26
5. Conclusions and Outlook 28
6. References 29

1. Introduction

Linear low-molecular-weight α-olefins have become very important industrial intermediates during the past decades. This is partly due to recent developments in homogeneous catalysis such as the Oxo and Wacker processes, and partly to increased demand for biodegradable detergents and for plasticizers and additives for rubber or gasoline.

Ethylene is a very attractive starting material for the synthesis of olefins and commercial production today is mainly by the Ziegler process, which utilizes the "Aufbau" reaction (1) where ethylene is oligomerized on aluminum alkyls at relatively high temperature (100–150° C) and pressure (60–150 atm):

Growth:

$$CH_3-CH_2-Al\begin{smallmatrix}R\\R\end{smallmatrix} + CH_2=CH_2 \rightarrow CH_3-CH_2-CH_2-CH_2-Al\begin{smallmatrix}R\\R\end{smallmatrix}$$

Displacement:

$$R-CH_2-CH_2-Al\begin{smallmatrix}R\\R\end{smallmatrix} \xrightarrow{slow} R-CH=CH_2 + H-Al\begin{smallmatrix}R\\R\end{smallmatrix}$$

$$H-Al\begin{smallmatrix}R\\R\end{smallmatrix} + CH_2=CH_2 \xrightarrow{fast} CH_3-CH_2-Al\begin{smallmatrix}R\\R\end{smallmatrix}$$

The metathesis reaction of low-molecular-weight olefins (C_2, C_3, C_4) aiming at the higher homologs (2), and the dehydrogenation of paraffins are other possible routes to the olefins, but these have not attained the same importance.

Transition-metal catalysts of the Ziegler-Natta type[1] (i.e. transition-metal compounds in combination with aluminum alkyls) were originally

---

[1] The original discovery of this type of catalysts is, of course, due to K. Ziegler, G. Natta having contributed ingenious applications, particularly to stereospecific polymerization. However we use the expression "Ziegler-Natta catalysts" in order to differentiate the bimetallic transition metal—aluminum alkyl catalysts from the monometallic aluminum alkyl catalyst of the "Aufbau" reaction, also due to Ziegler.

conceived for the production of high-molecular-weight polyethylene and other polyolefins. In recent years they have repeatedly been modified with the aim of producing oligomeric material (3–7). Transition-metal catalysis, its results and prospects, as well as its kinetics and mechanism, is the subject of this article.

Transition metals, especially nickel, also play a certain role in the "Aufbau" reaction by regulating the molecular weight (8), but the important difference between the "Aufbau" process and the catalysis reviewed in the present paper is that in the former chain growth takes place at the aluminum center. Wilke *et al.* (9) have suggested that the action of the nickel goes via a tris(ethylene)-Ni(O) complex that operates as a chain-transfer agent. With Ziegler-Natta type catalysts, on the other hand, the chain grows at the transition-metal center (10, 11). The aluminum alkyl may have a double task: to alkylate the transition-metal compound, and to form a complex with it, thus activating the transition metal—carbon bond. In certain cases decomposition of a metal—carbon bond may also cause reduction of the metal ion.

As a working hypothesis we shall assume the following general structure for the catalysts under consideration:

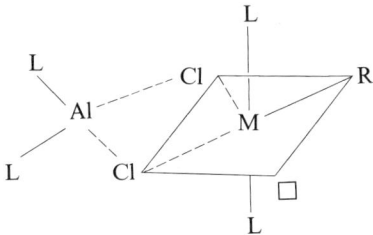

with M = transition metal, R = alkyl group or growing chain, L = variable ligands, and □ = vacant coordination site. The suggested structure — an octahedral arrangement of the ligands around the transition metal with chlorine bridges (three-center bonds) between M and aluminum — has been corroborated by magnetic and other proofs in the particular case of the catalyst system bis(π-cyclopentadienyl)-titanium-dichloride/aluminum alkyl, which polymerizes ethylene. Not only the ligands L at the transition-metal center, but also the exocyclic ligands at the aluminum have been shown to exert a remarkable influence on the stability (and hence reactivity) of the metal—carbon bond $M-R$; this underlines the importance of complex formation in Ziegler-Natta type catalysis (12, 13).

2. General Concepts

2.1. The β-Hydrogen Abstraction

The coordinative polymerization of olefins at a transition-metal center M is characterized by the presence of a metal—carbon bond able to insert a monomer molecule previously coordinated to the same metal center (chain-propagation step):

$$\begin{array}{c} H_2C\!=\!CH_2 \\ \vdots \\ M\!-\!CH_2\!-\!CH_2\!-\!R \end{array} \to M\!-\!CH_2\!-\!CH_2\!-\!CH_2\!-\!CH_2\!-\!R \,. \tag{1}$$

The principle regulating molecular weight is the β-hydrogen abstraction, a common reaction for breaking down transition metal—carbon bonds (14, 15). A hydrogen atom from the β-position of the alkyl group attached to the metal is transferred to the latter with formation of a metal hydride, and the organic residue leaves the metal center as a vinyl olefin. If this type of reaction occurs in a polymerizing system, it has to be considered from a kinetic point of view as a chain-transfer reaction, because the metal hydride with monomer will give an alkyl group able to continue the kinetic chain:

$$M\!-\!CH_2\!-\!CH_2\!-\!R \to MH + CH_2\!=\!CH\!-\!R \xrightarrow{+\,C_2H_4} M\!-\!CH_2\!-\!CH_3 \,. \tag{2}$$

The stability of the metal—alkyl bond toward β-hydrogen abstraction depends on the metal, its valency state and, very importantly, on the ligand environment. Conditions have been found which provide the transition metals of the left-hand end of the transition-metal series with a relatively high stability: Ti, V, Cr, Mo can build good polymerization catalysts, and these have found industrial application for the production of polyethylene and copolymers.

A different pattern is found with the right-hand end of the transition-metal series. These metals are much more prone to β-hydrogen abstraction from an attached alkyl group. This is why complexes based on these metals — e.g. Ni (16) or Rh (17) — are particularly well suited as catalysts for the dimerization and codimerization of olefins[2]. In the case of ethylene, the dimer butene-1 is formed by β-hydrogen transfer after the first irreversible propagation step:

$$M-H + CH_2=CH_2 \rightleftarrows M-CH_2-CH_3$$
$$M-CH_2-CH_3 + CH_2=CH_2 \rightleftarrows M-CH_2-CH_2-CH_2-CH_3$$
$$M-CH_2-CH_2-CH_2-CH_3 \longrightarrow M-H + CH_2=CH-CH_2-CH_3 .$$

[Most dimerization catalysts are able to act at the same time as isomerization catalysts, transforming butene-1 to cis- and trans-butene-2 (16)].

It therefore follows that there are two potential routes to the desired low-molecular-weight olefins: (1) a modification of polymerization catalysts (valency, ligands) appears attractive in that it could lower the stability of the metal—carbon bond toward β-hydrogen transfer to a level where, after 3 to 10 propagation steps, the alkyl chain would leave the metal center as an α-olefin; (2) alternatively, one could, when using dimerization catalysts, increase the stability of the metal—carbon bond to the stage where not just two but several ethylene units can grow before the α-olefin is split off.

Of these two routes only the first has been successful so far. Active oligomerization catalysts have been reported to produce linear olefins in the range of interest; they consist of Ti(IV) or Zr(IV) compounds in combination with aluminum alkyls (see Section 3).

2.2. Stability of the Metal toward Reduction

The lability of many transition metal—alkyl compounds toward reduction is a further feature in this type of catalysis. Formally, the reaction may be written:

$$M^n - R \rightarrow M^{n-1} + \cdot R , \qquad (3)$$

---

[2] This simple division into polymerization and dimerization catalysts does not apply if the growing chain stabilizes as a π-allyl—metal complex, as is the case with conjugated diolefins (18). Certain cobalt complexes, for instance, are perfectly able to polymerize butadiene (19).

which means that one of the two electrons forming the σ bond between M and R remains with the metal and the other with the alkyl group. Most Ti(IV)—alkyl compounds are reduced in this way at higher temperatures. The consequence is either complete depletion of catalytic activity, as in the case of the system bis(π–C_5H_5)–Ti(IV)–(C_2H_5)Cl/$C_2H_5AlCl_2$ (12, 13, 20), or the transformation of an oligomerization catalyst into a polymerization catalyst (see Section 3). In general, Zr(IV) catalysts are considerably less sensitive to reaction (3) than Ti(IV) compounds, thus permitting operation at higher temperatures.

Reaction (3) does not lead to free radicals, as might appear from the equation. The detailed reaction mechanism appears rather to be related to the β-hydrogen abstraction discussed in the preceding section. Two suggestions have been made. The first refers to the bis(cyclopentadienyl)-Ti(IV) system mentioned above. Having obtained an experimental second-order rate for the reduction of Ti(IV) to Ti(III), the present authors assumed the reaction to be bimolecular, with β-hydrogen transfer from one Ti unit to the other followed by donation of the hydrogen atom to the alkyl group of the second (formation of ethane) and liberation of ethylene at the first (13):

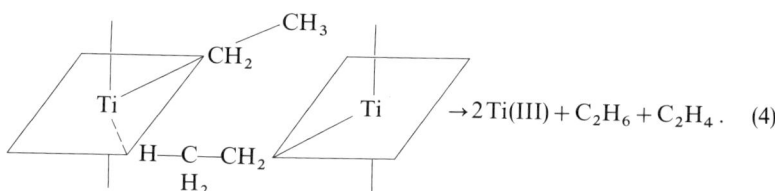

This mechanism has been corroborated by the isolation of a binuclear zirconium(IV) complex

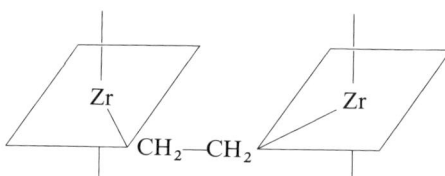

under similar reaction conditions [Sinn et al. (21)].

Whitesides et al. (22), on the other hand, suggested a two-step mechanism in the case of a Cu(I)-alkyl, namely

$$CH_3CH_2CH_2CH_2CuP(n\text{-}Bu)_3 \rightarrow CH_3CH_2CH{=}CH_2 + HCuP(n\text{-}Bu)_3 \quad (5a)$$

$$CH_3CH_2CH_2CH_2CuP(n\text{-}Bu)_3 + HCuP(n\text{-}Bu)_3 \rightarrow$$
$$CH_3CH_2CH_2CH_3 + 2\,Cu^\circ + 2\,P(n\text{-}Bu)_3 \quad (5b)$$

The formation of a copper hydride as an intermediate in the reaction was deduced from the presence of hydrogen among the products of thermal decomposition of the Cu(I) alkyl, and from the formation of additional hydrogen on acidification of the solution that remained after complete decomposition.

2.3. Copolymerization of α-Olefins (Branching)

The α-olefins formed during the oligomerization of ethylene are able to compete with ethylene for the coordination sites at the transition-metal center. Although coordinating ability falls with increasing molecular weight, the butene-1 accumulating in the reaction solution with higher conversion becomes a serious competitor for ethylene, especially at higher temperatures, where the solubility of ethylene in the reaction medium is relatively low.

The incorporation of α-olefins into the growing chain leads to branched molecules, and hence is a considerable nuisance; in most cases it limits the useful conversion to some 10–15 moles of ethylene per liter of reaction solution. Moreover, if β-hydrogen transfer takes place after such a copolymerization step, either unwanted vinylidene endgroups form or inner unsaturation occurs, depending upon whether the incorporation of the α-olefin took place according to Markownikoff's rule or was "anti-Markownikoff" (Eqs. (6) and (7), respectively, formulated for the incorporation of butene-1):

$$\text{Ti}-\text{R} + \text{CH}_2=\underset{\underset{\text{CH}_3}{|}}{\underset{|}{\text{CH}}}-\text{CH}_2 \rightarrow \text{Ti}-\text{CH}_2-\underset{\underset{\text{CH}_3}{|}}{\underset{|}{\text{CH}}}-\text{CH}_2-\text{R} \rightarrow \text{Ti}-\text{H} + \text{CH}_2=\underset{\underset{\text{CH}_3}{|}}{\underset{|}{\text{C}}}-\text{CH}_2-\text{R} \quad (6)$$

$$\text{Ti}-\text{R} + \underset{\underset{\text{CH}_3}{|}}{\underset{|}{\text{CH}}}-\text{CH}_2=\text{CH}_2 \rightarrow \text{Ti}-\underset{\underset{\text{CH}_3}{|}}{\underset{|}{\text{CH}}}-\text{CH}_2-\text{R} \rightarrow \text{Ti}-\text{H} + \underset{\underset{\text{CH}_3}{|}}{\underset{|}{\text{CH}}}=\text{CH}-\text{R} \quad (7)$$

Apart from the more trivial ways of keeping the conversion low and the ethylene concentration high, there could be two potential ways of preventing the α-olefins from incorporation, involving electronic and steric effects, respectively. The first is to take advantage of the higher coordination ability of ethylene, increasing the electron density at the metal center by systematically varying the ligands to the extent that, hopefully, only ethylene is able to be coordinated. The second is to try to ensure that the coordination site is blocked sterically by bulky ligands so that only the small ethylene molecule has access to the site. Both approaches have been used with moderate success (see Section 3).

3. Catalyst Tailoring in Ethylene Oligomerization

3.1. Ti—Al Catalysts

In 1959 Bestian et al. (*3, 23, 24*) first used a bimetallic catalyst system of the Ziegler-Natta type (TiCl$_4$ or an alkylated Ti(IV) compound in combination with an alkyl aluminumhalide) at extremely low temperatures (-100 to $-50°$) for the oligomerization of ethylene.

TiCl$_4$ with triethylaluminum is one of the "classic" Ziegler-Natta systems for the polymerization of ethylene to high-molecular-weight material. However, the soluble Ti(IV) compound is reduced to insoluble TiCl$_3$, even at low temperatures, and a heterogeneous Ti/Al surface complex is generally assumed to be the active species.

With monoalkyl-aluminum dichloride, on the other hand, no reduction occurs at room temperature and below. The catalyst remains in solution and in the presence of ethylene oligomer is formed. Evidently, the relatively low electron density at the Ti(IV) center (high electron affinity, high acidity, or however one wishes to express the situation) favors the molecular weight-reducing β-hydrogen abstraction, Eq. (2). Not only the valency of the titanium ion itself, but also the presence of the acceptor ligands Cl at the titanium center and at the aluminum alkyl contribute to the acidity of the catalyst center.

With their systems Bestian et al. obtained oligomers in the molecular-weight range 500–3000, but the oligomers were predominantly 2-ethyl-1-olefins, i.e. possessing vinylidenic endgroups. Table 1 gives the molecular-weight distribution and composition of a typical example, prepared in CH$_2$Cl$_2$ as solvent. Clearly, the low electron density at the metal center favors not only the β-hydrogen abstraction but also the coordination of the α-olefins, once formed, and their incorporation into the growing chain according to Eq. (6).

A somewhat different pattern is obtained if the catalytic system TiCl$_4$/C$_2$H$_5$AlCl$_2$ is used at 5° and in an aromatic solvent (*20*). Again,

Table 1. Oligomerization of ethylene with CH_3TiCl_3/CH_3AlCl_2 in CH_2Cl_2.
$T = -70°$; $[Ti] = 0.11$ mol/l; $[Al]/[Ti] = 1$ (3)

Fraction	Referred to converted C_2H_4 %	Composition
C_4	14	Butene-1
C_6	39	15% Hexene-1 78% 2-ethyl-butene-1
C_8	21	7,5% Octene-1 82% 2-ethyl-hexene-1
C_{10}	7	10% Decene-1 80% 2-ethyl-octene-1
$>C_{10}$	19	Mostly 2-ethyl-olefins

Table 2. Oligomerization of ethylene with $TiCl_4/C_2H_5AlCl_2$ in benzene. $T = +5°$; $[Ti] = 0.02$ mol/l; $[Al]/[Ti] = 5$; Ethylene Pressure: 1 atm; Conversion: 12 mol/l (20)

Weight fraction (%)	Molecular weight	Molecules bearing a double bond (%)			Degree of branching[a]
		Vinyl	Vinylidene	Inner	
6,5	190	15	16	69	0.7
13,5	230	8	9	83	1.3
20,3	310	5	4	91	2,0
12,5	400	2	5	93	2,6
21,5	540	2	5	93	3,7
15,5	570	2	6	92	3,4
10,5	700	—	—	—	—

[a] Branches per molecule.

the copolymerization is quite important but in this case mostly olefins with inner unsaturation are found. Table 2 shows fractionation data of an oligomer obtained with this system at a conversion of 12 mol/l. (Note that throughout this paper conversions are given in moles of ethylene converted per mole of original reaction solution, i.e. without correction for the increase in volume during the reaction). The lowest-molecular-weight part of the distribution which distills off with the solvent has not been investigated. The percentage of vinylic, vinylidenic and inner unsaturation has been determined by IR, the degree of branching by nmr. The predominant occurrence of inner double bonds appears to indicate that, under these reaction conditions, the α-olefins are incorporated preferentially in an "anti-Markownikoff" fashion, according to Eq. (7). This is surprising.

The two experiments represented in Tables 1 and 2 illustrate a general difficulty in homogeneous catalysis. The processes aimed at consist mostly of complicated reaction sequences, every step of which may have different, and sometimes opposite requirements as regards ligand influences. In the present case, if linear low-molecular-weight α-olefins are sought, some compromise between influences that lower molecular weight and those that reduce copolymerization of the α-olefins appears to be inevitable.

As already mentioned in Section 2.3, donor ligands are expected to diminish the incorporation of α-olefins. For a proper selection of donor ligands we may compare the ionization potentials of substituted benzenes. Good donor substituents lift the energy level of the highest occupied molecular orbital of the aromatic molecule, i.e. they provoke a decrease of the first ionization potential as compared with that of unsubstituted benzene (see Table 3).

Table 3. Ionization potentials of some substituted benzenes (25)

Substituent	I (eV)
—H	9.245
—CH_3	8.82
—C_2H_5	8.76
—OH	8.50
—SH	8.33
—C_6H_5	8.27
—OCH_3	8.22
—OC_2H_5	8.13

Ethoxy ligands (with the Ti compound) and ethyl ligands (with the aluminum alkyl) have been chosen to modify the catalyst system used for Table 2. Unfortunately, the most dramatic influence observed when acceptor ligands Cl are replaced by donor ligands is that on the molecular weight (Table 4). Whereas with $TiCl_4/C_2H_5AlCl_2$, 92% of the product was benzene-soluble oligomer, the insoluble high-molecular part increases up to 90% with $(C_2H_5O)TiCl_3/(C_2H_5)_2AlCl$. Interestingly, the substitution of one Cl by C_2H_5 at the aluminum has the same effect on the molecular weight as the substitution of four Cl by C_2H_5O at the titanium. (The absence of Ti (III) was checked by esr.) A similar observation was also made during the oligomerization of propene (26).

With the catalytic systems under consideration, contrarily to most polymerizing systems, the degree of polymerization has a positive

Table 4. Influence of the ligands on the molecular weight of the product. Reaction conditions as in Table 2. [(20); partly (28)]

Ti-Compound	Al-Compound	Oligomer (%)	Solid polymer (%)
$TiCl_4$	$C_2H_5AlCl_2$	92	8
$(C_2H_5O)TiCl_3$	$C_2H_5AlCl_2$	77	23
$(C_2H_5O)_3TiCl$	$C_2H_5AlCl_2$	56	44
$(C_2H_5O)_4Ti$	$C_2H_5AlCl_2$	31	69
$TiCl_4$	$(C_2H_5)_2AlCl$[a]	30	70
$(C_2H_5O)TiCl_3$	$(C_2H_5)_2AlCl$[a]	10	90

[a] [Al]/[Ti] = 2.

overall activation energy. In other words, the molecular weight decreases with decreasing temperature (see also Section 4.2). The most favorable compromise was found with the system $(C_2H_5O)_3TiCl/C_2H_5AlCl_2$ at −20°. An ethylene pressure of 12 atm additionally favored the coordination of ethylene as compared with that of the α-olefins. Table 5 presents fractionation data for an oligomer obtained with this system. The lower fractions are essentially linear α-olefins. Only the higher fractions show some branching, but the low content of vinylidene endgroups and of the inner unsaturation also demonstrate the small extension of the copolymerization reaction. The number-average molecular weight of the total oligomer corresponds to a degree of polymerization of 5.

Table 5. Oligomerization of ethylene with $(C_2H_5O)_3TiCl/C_2H_5AlCl_2$ in toluene. [Ti] = 0.02 mol/l; [Al]/[Ti] = 5; T = 20°; t = 1 h; ethylene pressure: 12 atm; conversion: 12 mol/l (5, 20)

Weight fraction (%)	Molecular weight	Molecules bearing a double bond (%)			Degree of branching
		Vinyl	Vinylidene	Inner	
9.8	56	100	0	0	0
12.9	84	100	0	0	0
14.8	112	95	5	0	0
12.8	171	90	7	3	0
22.3	213	87	9	4	0.2
12.4	292	70	20	10	0.4
14.2	465	70	20	10	1.3
0.8	800	—	—	—	—
Average	140	93	5	2	0.09

A similar optimization of a Ti-Al catalyst as regards catalyst composition, temperature, and ethylene pressure has been carried out by Langer (4) with the system $TiCl_4/t\text{-BuOH}/C_2H_5AlCl_2/(C_2H_5)_2AlCl$. From a kinetic evaluation of some of Langer's data (see Section 4.3) it may be concluded that very similar products are obtained at $-20°$.

3.2. Sulfur-Containing Additives

The effect of the addition of neutral donor ligands to the catalyst system was investigated by adding sulfur-containing compounds such as thioxanthene or thiols.

Like oxygen, bonded sulfur has lone-pair electrons that provide it with donor properties, but $d\pi$ orbitals are also available for use in π-bonding [this bonding seems to be even more prominent with sulfur than with phosphorus compounds (27)]. Both properties appear to predispose sulfur compounds to form ligands of transition-metal ions.

The addition of sulfur compounds to a solution of $TiCl_4$ in 1,2-dichloroethane or toluene causes a color change from yellow to brown, indicative of complex formation. The addition of aluminum alkyl produces a further deepening of the color but the solution remains transparent; no Ti(III) is formed. Although the structure of the active catalyst species is not yet known, it is reasonable to assume that all three components form part of it.

The sulfur-containing additive may be expected to have a dual influence, both electronic and steric. The increase in electron density at the transition-metal center may prevent coordination of the α-olefins; this effect must necessarily have a bearing on the coordination of the ethylene itself, decreasing the rate of ethylene consumption. On the other hand, very bulky sulfur ligands may block the coordination site effectively enough to make it accessible only for the small ethylene molecule, but not for the higher α-olefins.

Table 6 shows how the addition of thioxanthene influences the oligomerization of ethylene with the system $TiCl_4/C_2H_5AlCl_2$ in 1,2-dichloroethane. The incorporation of α-olefins is reduced, as shown by the decreased number of branches per molecule as well as by the distribution of double bonds. The most striking feature is perhaps the drastic diminution of the inner double bonds. Whereas in the absence of thioxanthene there is a clear tendency to the anti-Markownikoff addition of α-olefin (see also Table 2), the appearance of vinylidene endgroups in the presence of the sulfur compound indicates that, if α-olefins

Table 6. Oligomerization of ethylene with $TiCl_4/C_2H_5AlCl_2$ in the presence of thioxanthene (S). [Ti] = 0.03 mol/l; [Al]/[Ti]: 2; T = 0°; t = 2 h; ethylene pressure: 20 atm; solvent: 1,2-dichloroethane (28)

$\frac{[S]}{[Ti]}$	Conversion (mol/l)	Oligomer (%)	Polymer (%)	Molecules bearing a double bond (%)			Degree of branching
				Vinyl	Vinylidene	Inner	
0	103	99	1	0	2	98	3.9
1	75	98	2	34	29	37	1.5
2	59	91	9	57	40	3	1.2
4	33	80	20	55	42	3	1.2

are incorporated under these conditions, it is according to the Markownikoff mode.

As expected, the overall rate is decreased in the presence of the additive, whereas the molecular weight does not appear to be greatly influenced. The addition of more than 2 moles of sulfur compound per mole of Ti-compound does not further improve the result.

An improvement may, however, be obtained by changing the solvent from 1,2-dichloroethane to toluene (see Table 7, first run). Table 7 also shows the effect of thioxanthene compared with that of other sulfur-containing compounds. Whereas the electronic effect may be assumed to be roughly the same because of the similar overall rates, the very bulky thioxanthene ligand appears to have a steric influence as well, thus preventing α-olefins from coordinating.

The oligomeric part of the product of the first run in Table 7 has been fractionated by vacuum distillation (Table 8). With the exception of the absence of inner double bonds, the pattern of unsaturation and branching is not very different from that obtained with the $(C_2H_5O)_3TiCl$

Table 7. Influence of different sulfur ligands. Catalyst system: $TiCl_4/C_2H_5AlCl_2$; [Ti] = 0.03 mol/l; [Al]/[Ti] = 2; [S]/[Ti] = 2; T = 5°; t = 4 h; ethylene pressure: 20 atm; solvent: toluene (28)

Sulfur ligand	Conversion (mol/l)	Oligomer (%)	Molecules bearing a Double Bond (%)			Degree of branching
			Vinyl	Vinylidene	Inner	
Thioxanthene	43	89	80	20	—	0.46
Dodecanethiol	46	98	70	30	—	1.80
Ethyl-thiophen-2-carboxylate	41	64	70	20	10	1.40

Table 8. Fractionation data for the first run in Table 7 (28)

Weight fraction (%)	Molecular weight	Molecules bearing a Double Bond (%)			Degree of branching
		Vinyl	Vinylidene	Inner	
0.96	56	100	—	—	—
10.55	84	95	5	—	—
16.16	112	95	5	—	0.1
9.56	177	83	17	—	0.2
10.16	192	81	19	—	0.2
11.07	206	80	20	—	0.3
9.86	242	78	22	—	0.5
8.66	263	77	23	—	0.5
13.46	312	76	24	—	0.5
9.56	440	63	37	—	1.5

system shown in Table 5. This result is remarkable, since it was obtained at a conversion of 43.6 moles of ethylene per liter, i.e. at a considerably higher concentration of competing α-olefins than in the case of $(C_2H_5O)_3TiCl$, where the conversion was only 13 mol/l. Furthermore, the $TiCl_4$/thioxanthene system can be operated at 0° (versus −20° in Table 5).

The remarkable coordination ability of thioxanthene and its blocking effect are assumed to be related to the folded structure of the compound, which has a dihedral angle of 135° (29):

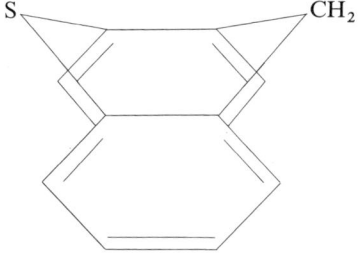

3.3. Zr—Al Catalysts

A most interesting recent development comprises the use of zirconium-based Ziegler-Natta catalysts for the oligomerization of ethylene. A

Montecatini patent (6) claims the production of olefins with tetrapropoxyzirconium in combination with one of several aluminum alkyls at 80° in heptane or toluene solution. Evidently the intermediate Zr alkyls (alkyl = ethyl or growing chain) are considerably more stable toward reduction of the metal than the corresponding titanium alkyls, thus permitting operation at higher temperatures, with the advantage of a faster reaction rate. Three runs carried out with the same aluminum alkyl (ethylaluminum sesquichloride), but at varying ethylene pressure, are summarized in Table 9. (These data will be used in Section 4.3. for a kinetic evaluation). The product was analyzed by gas chromatography and is claimed to consist essentially of linear α-olefins.

A somewhat different Zr/Al catalyst has been reported by Attridge et al. (7): tetrabenzyl zirconium in combination with ethylaluminum sesquichloride has been shown to oligomerize ethylene at 40–80° in toluene solution. From analyses of the product by gas-liquid chromatography, infrared and nmr spectroscopy, it is concluded also in this case that the product is essentially pure α-olefins. The catalyst system is reported to have a higher activity (higher reaction rate at comparable

Table 9. Oligomerization of ethylene with $Zr(O-nC_3H_7)_4/Al_2(C_2H_5)_3Cl_3$. $T = 80°$ (6)

Run No.[a]	I	II	VIII
$[Zr] \times 10^3$ (mol/l)	1.3	4.0	2.3
[Al]/[Ti]	9	9	10
Solvent	Heptane	Heptane	Toluene
$P_{C_2H_4}$ (atm)	25	45	10
t (min)	30	60	30
Conversion (mol/l)	18	45	10
Distribution [Weight (%)]			
C_4	5.1	0.7	8.2
C_6	9.2	6.0	7.8
C_8	9.1	8.6	5.4
C_{10}	9.3	11.3	12.4
C_{12}	10.8	12.1	12.2
C_{14}	11.6	11.6	12.0
C_{16}	9.6	10.3	9.4
C_{18}	9.1	9.7	7.6
C_{20}	7.7	8.8	7.1
C_{22}	6.3	7.1	5.7
C_{24}	4.5	4.5	4.5
C_{26}	3.4	3.9	3.4
C_{28}	—	2.4	2.7
C_{30}	—	1.9	1.6

[a] Refers to the examples in the original patent.

conditions) than the Montecatini system. The difference resides in the ligands of the Zr(IV) center. Presumably the zirconium—carbon bonds in tetrabenzyl zirconium are able to initiate the oligomerization (after complex formation Zr-Al), whereas in the case of tetrapropoxy zirconium the active species must be formed, replacing at least one propoxy group by an ethyl group, which might be a slow equilibrium process.

3.4. Aluminum-free Ti Catalysts

For the sake of completeness, a slow oligomerization of ethylene with methyltitanium dichloride, in the absence of an aluminum alkyl, should be mentioned. This reaction was reported by Kühlein and Clauss (30); it takes place at $-70°$ in CH_2Cl_2 solution at normal pressure. A very slow growth reaction (some 5–10 growth steps per titanium center and per hour) is occasionally interrupted by a β-hydrogen abstraction. After some hours, a mixture of olefins and paraffins (the latter proceeding from growing chains, quenched with methanol) in the range C_3–C_{30} may be obtained. The reaction is interesting, because it shows that in principle chain growth at a monometallic (and mononuclear) titanium center is possible; however it again points to the importance of complex formation between transition metal and aluminum alkyl in the Ziegler-Natta type catalysts, where an activation of the transition metal—carbon bond ensures satisfactory reaction rates.

4. Kinetics and Mechanism

4.1. Kinetic Equations and Conditions for their Application

Several of the catalytic systems reported in Section 3 have been shown to give, under certain restricted reaction conditions, essentially (or predominantly) linear α-olefins. In this case, and in the absence of chain termination reactions (constant overall rate), only two reactions have to be considered in the reaction scheme: chain propagation and chain transfer [cf. Eqs. (1) and (2)].

The rate of chain propagation r_p may be formulated as

$$r_p = k_p[P^*][M] \qquad (8)$$

where $[P^*]$ and $[M]$ are the concentrations of active sites and of monomer, respectively. This rate law is generally assumed to apply to soluble Ziegler-type catalysts.

The rate of chain transfer may be either

or
$$r_{tr} = k_{tr}[P^*][M] \qquad (9a)$$
$$r_{tr} = k_{tr}[P^*], \qquad (9b)$$

depending on whether the monomer is involved in the rate-determining step of the chain transfer [cf. Eq. (2)]. Evidently, the overall kinetics depend very heavily upon this mechanistic feature.

Where Eq. (9a) is valid, monomer is used up not only by chain propagation but also in the transfer step. Hence, the overall rate of monomer consumption is

$$-d[M]/dt = (k_p + k_{tr})[P^*][M] \qquad (10a)$$

and the number-average degree of polymerization is

$$P_n = \frac{r_p + r_{tr}}{r_{tr}} = \frac{k_p}{k_{tr}} + 1, \qquad (11a)$$

If Eq. (9b) is valid:

$$-d[M]/dt = k_p[P^*][M] \qquad (10b)$$
$$P_n = r_p/r_{tr} = k_p[M]/k_{tr}. \qquad (11b)$$

If the average molecular weight of the total oligomer is experimentally available, its monomer dependence allows one to decide which of the two reaction schemes (set a or set b) is applicable. In either case, a "normal" molecular weight distribution (Schulz-Flory distribution) results. This means that the weight fraction m_P of oligomer of degree of polymerization P is given by (18, 31):

$$m_P = \ln^2 \alpha \cdot P \cdot \alpha^P \tag{12}$$

where α is the probability of the propagation step:

$$\alpha = \frac{r_p}{r_p + r_{tr}} = \frac{1}{1 + \frac{r_{tr}}{r_p}}. \tag{13}$$

From Eq. (12) it follows that

$$\log(m_P/P) = \log(\ln^2 \alpha) + P \cdot \log \alpha. \tag{14}$$

A plot of log (m_P/P) versus P permits the determination of log α as the slope of a straight line. Hence, in cases where chromatographic fractionation data are available, α may be determined graphically. This is also true if low-molecular-weight material has been lost during working up, provided that the portion of the experimentally reliable fraction, say C_{10}–C_{24}, gives a linear plot.

Introduction of Eqs. (8) and (9a) into (13) gives

$$\alpha/(1-\alpha) = k_p/k_{tr} \tag{15a}$$

whereas application of (9b) results in

$$\alpha/(1-\alpha) = k_P[M]/k_{tr}. \tag{15b}$$

An evaluation of the molecular-weight distribution (or parts thereof) thus provides an independent way of distinguishing between the two proposed reaction schemes. Once this problem is settled, the individual rate constants k_p and k_{tr} become accessible from the data on overall rate and molecular weight [Eqs. (10) and (11)] or from data for overall rate and fractionation [Eqs. (10) and (15)]. The a or b set is used as appropriate but, of course, the concentration of active sites $[P^*]$ and concentration of monomer $[M]$ must be known with sufficient reliability.

4.2. Evaluation of a Ti—Al System

The catalyst system $(C_2H_5O)_3TiCl/C_2H_5AlCl_2$ in toluene produces essentially linear α-olefins at temperatures of $-20°$ or below and at an ethylene pressure of 12 atm, provided the conversion is maintained below 10–15 moles ethylene/l of catalyst solution (Section 3.1; see also Refs. (5) and (20)). Hence, the kinetic equations of Section 4.1 are in principle applicable to the oligomerization of ethylene with this catalyst.

Molecular weight and fractionation data have to be consulted in order to decide for set a or set b. The exact determination of the average molecular weight of an oligomer is always complicated by the possible loss of volatile parts. Whereas, with some care, hexene and octane may be recovered more or less quantitatively, some loss of butene appears inevitable. By applying Eq. (12) one can correct for this loss. From this equation it follows that α may be determined from the weight fractions of two successive degrees of polymerization as

$$\alpha = \frac{P_n m_{P,n+1}}{P_{n+1} m_{P,n}}. \tag{16}$$

The weight fraction of butene may then be calculated directly from Eq. (12), with $P = 2$.

In the present case the following procedure was adopted: after reaction the autoclave is vented at $-20°$ (no loss of oligomer $\geqslant C_6$). The reaction mixture is weighed and washed with ice-water to separate the catalyst. The organic liquid phase is distilled at room temperature in vacuo. The distillate, containing the fractions $\leqslant C_{10}$ and the solvent, is analyzed quantitatively by gas chromatography, a known amount of C_7 being used as internal standard. The weight fractions of C_6 and C_8 serve to determine α. The residue containing the higher fractions is weighed, and its molecular weight is determined by vapor-pressure osmometry. Finally, the number-average molecular weight M_n of the total product is obtained from $M_n = 1/\Sigma\, m_i/M_i$. (The degree of polymerization one has not been taken into account; see, however, footnote 3.)

Molecular-weight data for several runs, at $-45°$ and at $-20°$, are given in Table 10. In contrast to the behavior of most polymerizing systems, the molecular weight increases with increasing temperature. Table 10 also shows that a variation of the monomer concentration (pressure) by a factor of four at $-20°$ does not cause an equivalent change in the molecular weight. Hence it may safely be concluded from the data that the degree of polymerization is not proportional to the monomer

Table 10. Ethylene oligomerization with $(C_2H_5O)_3TiCl/C_2H_5AlCl_2$: molecular weights. Solvent: toluene

Run No.	T (°C)	$[Ti] \times 10^3$ (mol/l)	[Al]/[Ti]	$P_{C_2H_4}$ (atm)	t (h)	Conversion (mol/l)	M_n
1	−45	20	7	12	1	4.8	90
2	−45	20	7	12	4	15.3	105
3	−20	10	7	12	1	9.0	138
4	−20	20	7	12	1	17.8	153
5	−20	20	5	12	1	13.0	140
6	−20	20	5	6	1	7.6	138
7	−20	20	5	3	4	13.5	201

Table 11. Ethylene oligomerization with $(C_2H_5O)_3TiCl/C_2H_5AlCl_2$: ratio of rate constants, k_p/k_{tr}. Ethylene pressure: 12 atm (see Table 10 for experimental details)

Run No.	T (°C)	α	k_p/k_{tr} [Eq. (11a)]	[Eq. (15a)]	Average
1	−45	0.74	2.2	2.8	2.8
2	−45	0.77	2.7	3.4	
3	−20	0.82	3.9	4.5	
4	−20	0.85	4.5	5.7	4.5
5	−20	0.82	4.0	4.5	

concentration [although M_n at 6 and 3 atm is certainly somewhat influenced by branching (20)]. Consequently, Eq. (11a) is the correct one for determining the rate constant ratio k_p/k_{tr}. The data are summarized in Table 11. (Only runs at $p = 12$ atm are taken into account.)

Although no complete fractionation data are available for the present system, the α-values obtained with Eq. (16) from the C_6 and C_8 fractions may be used to calculate the rate constant ratio from Eq. (15a) too. These data are included in Table 11. Concordance with the values obtained from Eq. (11a) is to be expected only in the case of a "Schulz-Flory" distribution of molecular weights. The relatively good agreement between the two sets of data in Table 11 hence corroborates the simple reaction scheme assumed in Section 4.1.

The overall rate of oligomerization gives some further information. Fig. 1 shows its dependence on ethylene pressure. Although there is a small amount of copolymerization (branching) at 6 atm, and every molecule has, on average, one branch at 3 atm (28), Figure 1 confirms

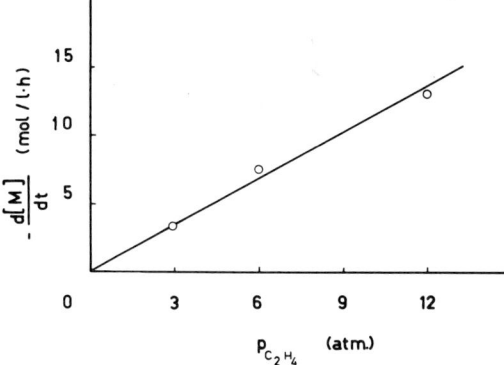

Fig. 1. Oligomerization of ethylene with $(C_2H_5O)_3TiCl/C_2H_5AlCl_2$ in toluene. Pressure dependence of the overall rate. $[Ti] = 20.10^{-3}$ mol/l; $[Al]/[Ti] = 5$: $T = -20°$ (28)

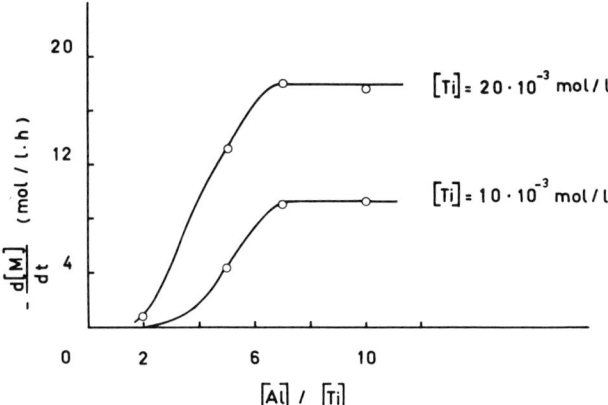

Fig. 2. Dependence of the overall rate on the ratio $[Al]/[Ti]$. $T = -20°$; ethylene pressure: 12 atm (32)

satisfactorily the validity of the rate low, Eq. (10a), as far as proportionality between the rate and the monomer concentration is concerned.

The overall rate depends also on the ratio of aluminum alkyl to titanium compound $[Al]/[Ti]$, increasing in the range up to $[Al]/[Ti] = 7$, and then levelling off (Fig. 2). At $[Al]/[Ti] \geqslant 7$, the rate is proportional to the titanium concentration.

Table 12. Ethylene oligomerization with $(C_2H_5O)_3TiCl/C_2H_5AlCl$ in toluene: rate data. Ethylene pressure: 12 atm

Run No.	T (°C)	$[Ti] \times 10^3$ (mol/l)	$\frac{[Al]}{[Ti]}$	Conversion (mol/l)	$\frac{d[M]}{dt} \times 10^3$ (mol/l · sec)	$(k_p+k_{tr}) \times 10^2$ (l/mol·sec) [Eq. (10a)]	Average
1	−45	20	7	4.8	1.3	1.2	
2	−45	20	7	15.3	1.1	1.0	1.1
3	−20	10	7	9.0	2.5	7.4	
4	−20	20	7	17.8	4.9	7.2	
8	−20	10	10	9.3	2.8	7.6	7.3
9	−20	20	10	17.5	4.9	7.2	

These data permit the assumption that the active centers are formed from the two components, titanium compound and aluminum alkyl, in an equilibrium reaction and that at $[Al]/[Ti] \geqslant 7$ essentially all titanium is present in the active form, i.e. $[P^*] = [Ti]_{total}$. The ethylene concentration in the reaction solution [M] is determined by the solubility of ethylene in toluene at the reaction temperature and pressure.

In the present case, literature data (33) have been extrapolated with the aid of Henry's law. The following values have been used: $[M]_{-20°} = 3.4$ mol/l, $[M]_{-45°} = 5.3$ mol/l.

The sum of the rate constants $(k_p + k_{tr})$ calculated from the rate law (10a) is given in Table 12 for the two temperatures under consideration. Only runs with $[Al]/[Ti] \geqslant 7$ have been evaluated, since only under this condition does $[P^*] = [Ti]_{total}$. Runs 1 and 2 indicate that $[P^*]$ may be assumed to be approximately constant during the reaction time. This is in agreement with the observation that no Ti(III) can be detected by esr (no catalyst depletion by reduction Ti(IV) → Ti(III), as in similar Ti/Al systems [12]).

The data given in Tables 11 and 12 finally permit us to determine the rate constants k_p and k_{tr} separately. Table 13 gives this information. The Arrhenius parameters are also included in the table. Though no more than rough estimates, these data indicate some trends.

The unusual temperature dependence of the molecular weight (smaller M_n with decreasing temperature) is traced back to the fact that $E_{tr} < E_p$. The low activation energy of the β-hydrogen abstraction

Table 13. The separated rate constants and their temperature dependence

T (°C)	$k_p \times 10^2$ (l/mol·sec)	E_p (kcal/mol)	$\log A_p$	$k_{tr} \times 10^2$ (l/mol·sec)	E_{tr} kcal/mol	$\log A_{tr}$
−45	0.8 }	9.2	6.7	0.3 }	6.7	3.9
−20	6.0			1.3		

is, in fact, amazing as compared with that of similar reactions:

$$\begin{array}{c} \text{Ti} \;\; \text{H} \\ |\;\;\;\;| \\ -\text{C}-\text{C}- \end{array} \longrightarrow \text{TiH} + -\text{C}=\text{C}- \quad\quad \begin{array}{l} 6\text{ kcal/mol} \\ \log A = 3.1 \end{array}$$

$$\begin{array}{c} \text{Al} \;\; \text{H} \\ |\;\;\;\;| \\ -\text{C}-\text{C}- \end{array} \longrightarrow \text{AlH} + -\text{C}=\text{C}- \quad\quad \begin{array}{l} 20\text{--}30\text{ kcal/mol } (34) \\ \log A = 10\text{--}12 \end{array}$$

$$\begin{array}{c} \text{Cl} \;\; \text{H} \\ |\;\;\;\;| \\ -\text{C}-\text{C}- \end{array} \longrightarrow \text{ClH} + -\text{C}=\text{C}- \quad\quad \begin{array}{l} 50\text{--}53\text{ kcal/mol } (35) \\ \log A = 13\text{--}14. \end{array}$$

Presumably, in the present case the alkyl group (growing chain) is highly polarized even in the ground state under the influence of the transition metal:

$$\begin{array}{c} \text{Ti}^{\oplus} \;\; \text{H}^{\ominus} \\ |\;\;\;\;\;\;| \\ \text{H}_2\text{C}^{\ominus}-\text{CH}^{\oplus}-\text{R}. \end{array}$$

Moreover, the extremely low A factor, as well as the occurrence of [M] in the rate law Eq. (9a), indicate a mechanism different from the one that leads to HCl and AlH in the reactions mentioned above.

It is suggested that a six-center, bicyclic, highly polar transition state including the monomer could account for the unusual activation parameters as well as for the rate law Eq. (9a):

$$H_2C=CH_2 \atop Ti-CH_2-CH_2-R \quad \rightarrow \quad {H_2C^\ominus \cdots CH_2^\oplus \atop Ti^\oplus \cdots H^\ominus \atop H_2C^\ominus \cdots CH-R} \quad \rightarrow \quad {H_2C-CH_3 \atop Ti} + CH_2=CH-R.$$

The propagation step, on the other hand, is assumed to proceed via a normal four-center transition state:

$$H_2C=CH_2 \atop Ti-CH_2-CH_2-R \quad \rightarrow \quad {H_2C^\ominus \cdots CH_2^\oplus \atop Ti^\oplus \cdots CH_2-CH_2-R} \quad \rightarrow \quad {CH_2-CH_2 \atop Ti \quad CH_2-CH_2-R}.$$

The mechanism suggested for the chain-transfer step is distinctly different from that of other β-hydrogen abstraction reactions from transition-metal alkyls. Several such reactions have been reported (15); they evidently proceed without the assistance of coordinated monomer.

4.3. Comparison with other Systems

It may reasonably be assumed that the mechanism deduced in the preceding section for the catalyst $(C_2H_5O)_3TiCl/C_2H_5AlCl_2$, is also valid for other relevant systems.

Data published by Langer (4) permit a pertinent estimate. The author used the catalyst system $(C_2H_5)_2AlCl/(C_2H_5)AlCl_2/TiCl_4$/t-BuOH to oligomerize ethylene. For a molar ratio 6:6:1:1 of the four catalyst components, an ethylene pressure of ca. 11 atm, and with chlorobenzene as solvent, sufficient data are available to estimate the rate constants at $-20°$, along the same lines as presented in Section 4.2. Under the given conditions, the oligomer is reported to consist essentially of linear α-olefins, having a normal molecular weight distribution. It has to be supposed, however, that all (or most) of the titanium is active. This means that the values of the estimated rate constants will be the lower

Table 14. Comparison of the rate constants of ethylene oligomerization with different catalyst systems

System[a]	T (°C)	$p_{C_2H_4}$ (atm)	$\dfrac{k_p}{k_{tr}}$	k_p (l/mol·sec)	k_{tr} (l/mol·sec)	α	C_6–C_{16} (%) [Eq. (17)]	Ref.
A	−45	12	2.8	0.008	0.003	0.74	55	Section 4.2
A	−20	12	4.5	0.060	0.013	0.82	42	Section 4.2
B	−20	11	2.9	0.074	0.019	0.75	54	Estimated from (4)
C	+80	10	3.5	1.9	0.56	0.78	50	Estimated from (6)
D	+40	11	2.2	3.7	1.4	0.69	59	Estimated from (7)

[a] A: $(C_2H_5O)_3TiCl/C_2H_5AlCl_2$ in toluene. B: $TiCl_4/t\text{-}BuOH/(C_2H_5)_3Al_2Cl_3$ in chlorobenzene; $[Ti] = 2 \times 10^{-3}$ mol/l; $r_p = 5.5 \times 10^{-4}$ mol/l·sec; $M_n = 109$; $k_p/k_{tr} = P_n − 1 = 2.9$. C: $(C_3H_7O)_4Zr/(C_2H_5)_3Al_2Cl_3$ in toluene. D: $Bz_4Zr/(C_2H_5)_3Al_2Cl_3$ in toluene (Bz = benzyl).

limits. The data are given in Table 14 (System B). The similarity of the two titanium systems is striking. Presumably the alkoxy groups form part of the active sites in Langer's system, too.

Moreover, Table 14 contains the result of an evaluation for the system $Zr(OC_3H_7)_4/(C_2H_5)_3Al_2Cl_3$. This system was reported (6) to produce linear α-olefins (cf. Section 3.3). The fractionation data given in Table 9 represent a satisfactory support for the presently suggested mechanism of β-hydrogen abstraction. These data are represented in Fig. 3, according to Eq. (14). The monomer concentration varies by a factor of 4.5 within the three runs. The close similarity of the measured values in the range C_{10}–C_{24} excludes a monomer dependence of α according to Eq. (15b), and hence corroborates the validity of the a set of equations, i.e. the participation of the monomer in the rate-determining step of the chain transfer.

With the use of Eq. (15a), a value of $k_p/k_{tr} \simeq 3.5$ may be estimated from the slope of the straight line in Fig. 3. Table 14 also contains a rough estimate (lower limits) of the separate rate constants k_p and k_{tr}, as obtained from the run at 10 atm in toluene (System C).

Finally, the Zr-Al system reported recently by Attridge et al. (7) is also included in Table 14 (System D). A somewhat different Zr/Al catalyst system has been used to oligomerize ethylene to linear α-olefins. Again the data permit a rough estimate of the rate constants along the same lines and with the same assumptions. From chromatographic

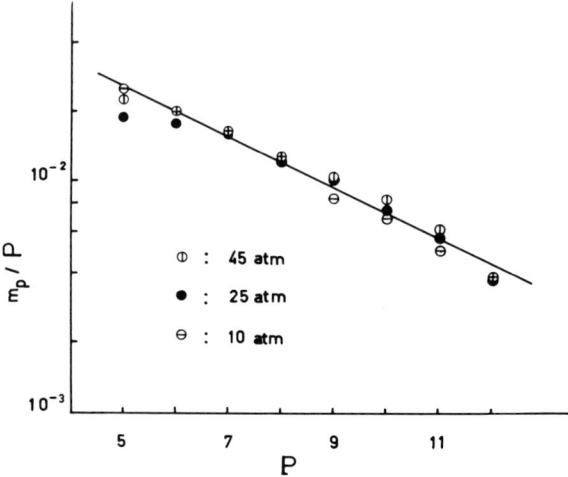

Fig. 3. Literature data on oligomerization of ethylene with Zr catalysts [6]; molecular-weight distribution at different ethylene pressure. $Zr(OC_3H_5)_4/(C_2H_5)_3AlCl_3$, T = 80°, heptane (●, ⊕) or toluene (⊖)

fractionation data given in the paper, it follows that the oligomer has a normal distribution with $\alpha = 0.69$.

The reaction conditions worked out for each of the systems are those where the catalyst behaves particularly well as an oligomerization catalyst. For titanium systems a temperature of $-20°$ or below is required, whereas zirconium systems work best at considerably higher temperatures. Although the individual rate constants are very different for the different systems — in particular some two orders of magnitude higher for the Zr systems than for the Ti catalysts — the ratio k_p/k_{tr} is similar in all cases.

4.4. Consequences of the Reaction Mechanism

The catalyst systems so far reported for the oligomerization of ethylene to linear α-olefins appear to operate all according to the same reaction mechanism which, in the absence of disturbing side reactions, leads to a Schulz-Flory distribution of molecular weights. This fixes the optimal weight fraction of oligomer of a particular molecular weight

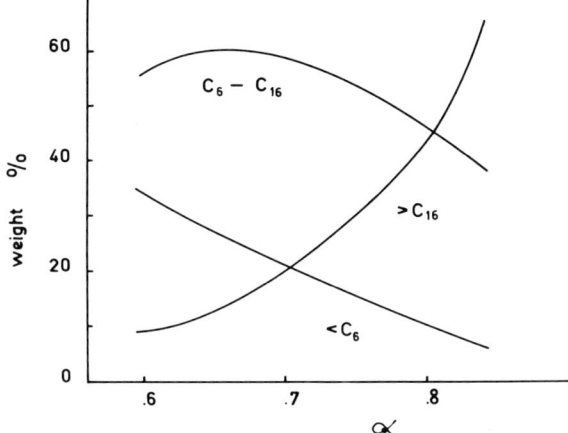

Fig. 4. Weight fractions of C_6—C_{16}, of $<C_6$ and of $>C_{16}$ as a function of α; calculated from Eq. (17)

range, say P_n to P_m, as [cf. Eq. (12)]:

$$\sum_{P=P_n}^{P_m} m_P = \sum_{P=P_n}^{P_m} ln^2\alpha \cdot P \cdot \alpha^P . \qquad (17)$$

The weight fraction of the technically most interesting range C_6–C_{16}, as calculated from this equation, is included in Table 14 for the systems treated there. Fig. 4 shows the relationship between this weight fraction and α. It is clear that the maximum attainable value for the range C_6–C_{16} is 60% (by weight) at $\alpha = 0.65$[3].

Table 14 indicates that with some of the systems this value has been approached (although not reached). It should be noted that, since α depends only on rate constants [Eq. (15a)], the temperature is the only variable by means of which a particular system may be adjusted to its optimum to give a certain desired molecular weight range.

---

[3] Equation (12) takes into account also those molecules that have been subject to chain transfer after the first incorporation of a monomer. Although these ethylene molecules do not contribute to the experimental overall conversion, they are the product of catalytic steps. For practical use it may be more convenient to refer the weight fraction of olefin in a certain range to the amount of ethylene converted to higher homologues. That means to normalize the weight fractions so that $\sum_{P=2}^{\infty} m_P$ is equal to unity. Under these conditions the maximum of C_6–C_{16} is found at $\alpha = 0.64$, with 67.5%.

Nevertheless, the ethylene pressure plays an important role in ethylene oligomerization. Not only does the overall rate increase linearly with $p_{C_2H_4}$ (Fig. 1), but the competition of α-olefin molecules for catalyst sites (branching) is also reduced at higher ethylene pressure.

5. Conclusions and Outlook

The problem of oligomerizing ethylene to low-molecular-weight, linear α-olefins with Ziegler-Natta type catalysts appears to be solved in principle. There remain, however, two more or less severe restrictions that limit the useful yield of oligomer.

The first restriction concerns the useful molecular weight range, and is inherent to the reaction mechanism. In the best of the cases (without disturbing side reactions) the reaction scheme is made up of reactions (1) and (2) (see Section 2.1) and leads then to a Schulz-Flory distribution of molecular weights, with the consequences outlined in Section 4.4 (cf. Fig. 4).

The second restriction concerns the conversion and is caused by the α-olefins formed competing with the ethylene for the coordination sites (branching, vinylidenic and inner unsaturation). So far all work reported has been done in batch experiments. A certain improvement might occur if the process could be converted into a continuous one. This would require (1) that the homogeneous catalysts (particularly the highly active zirconium compositions) be anchored on a carrier, and (2) that the reaction solution be replaced after a relatively short residence time over the catalyst. Although the "heterogenization" of homogeneous catalysts appears to be, at present, one of the prevailing themes in the field of transition-metal catalysis, no such endeavor has been reported so far for oligomerization of ethylene.

6. References

1. Ziegler, K.: Brennst.-Chem. **35**, 321 (1954).
2. Phillips Petroleum Company, US Patent 3 491 163; January 20, 1970.
3. Bestian, H., Clauss, K.: Angew. Chem. **75**, 1068 (1963).
4. Langer, A. W.: J. Macromol. Sci. Chem. A. **4**, 775 (1970).
5. Henrici-Olivé, G., Olivé, S.: Angew. Chem. **82**, 255 (1970); Angew. Chem. Intern. Edit. Engl. **9**, 243 (1970).
6. Montecatini Edison S.p.A., Netherlands Patent Specn. 70, 13193/1971.
7. Attridge, C. J., Jackson, R., Maddock, S. J., Thompson, D. T.: Chem. Commun. **1973**, 132.
8. Ziegler, K., Holzkamp, E., Breil, H., Martin, H.: Angew. Chem. **67**, 541 (1955).
9. Fischer, K., Jonas, K., Wilke, G.: Angew. Chem. **85**, 620 (1973); Angew. Chem. Intern. Edit. Engl. **12**, 565 (1973).
10. Boor, J.: Ind. Eng. Chem. Prod. Res. Develop. **9**, 437 (1970), and references therein.
11. Henrici-Olivé, G., Olivé, S.: J. Polymer Sci. C (Polymer Letters) **8**, 21 (1970).
12. – – Angew. Chem. **79**, 764 (1967); Angew. Chem. Intern. Edit. Engl. **6**, 790 (1967).
13. – – Advan. Polymer Sci. **6**, 421 (1969).
14. Mowart, W., Shortland, A., Yagupsky, G., Hill, N. J., Yagupsky, M., Wilkinson, G.: J. Chem. Soc. Dalton **1972**, 533.
15. Braterman, P. S., Cross, R. J.: J. Chem. Soc. Dalton **1972**, 657, and references therein.
16. Bogdanovic, B., Wilke, G.: Proceedings of the 7th World Petroleum Congress, Mexico 1967, p. 351.
17. Cramer, R.: J. Am. Chem. Soc. **87**, 4717 (1965).
18. See, e.g., Henrici-Olivé, G., Olivé, S.: Polymerisation – Katalyse, Kinetik, Mechanismen. Weinheim: Verlag Chemie 1969; and references therein.
19. Longiave, C., Castelli, R., Croce, G. F.: Chim. Ind. (Milano) **43**, 625 (1961).
20. Henrici-Olivé, G., Olivé, S.: Chem.-Ing.-Techn. **43**, 906 (1971).
21. Heins, E., Hinck, H., Kaminsky, W., Oppermann, G., Raulinat, P., Sinn, H.: Makromol. Chem. **134**, 1 (1970).
22. Whitesides, G. M., Stedronski, E. R., Casey, C. P., San Fillipo, J.: J. Am. Chem. Soc. **92**, 1426 (1970).
23. US Patent 2 907 805 (October 6, 1959). Bestian, H., Prinz, E.: Farbwerke Hoechst.
24. Bostian, H., Claus, K., Jensen, H., Prinz, E.: Angew. Chem. Intern. Edit. Engl. **2**, 32 (1963).
25. Streitwieser, A.: Progr. Phys. Org. Chem. **1**, 1 (1963).
26. Henrici-Olivé, G., Olivé, S.: J. Polymer Sci., Part B **8**, 205 (1970).
27. Cotton, F. A., Wilkinson, G.: Advanced Inorganic Chemistry (Second Edition). New York: Interscience Publ. 1966.
28. Unpublished work.
29. Aroney, M. J., Hoskins, G. M., Le Fevre, R. J. W.: J. Chem. Soc. (B) **1969**, 980.
30. Kühlein, K., Clauss, K.: Makromol. Chem. **155**, 145 (1972).
31. Schulz, G. V.: Z. Physik. Chem. B-**43**, 25 (1939).
32. Henrici-Olivé, G., Olivé, S.: J. Polymer Sci. (Polymer Letters) **12**, 39 (1974).

33. Waters,J.A., Mortimer,G.A., Clements,H.E.: J. Chem. Eng. Data **15**, 174 (1970).
34. Cocks,A.T., Egger,K.W.: J. Chem. Soc., Faraday I, **1972**, 423 (Vol. 68).
35. Egger,K.W., Cocks,A.T.: In: Patai,S. (Ed.): The Chemistry of the Carbon-Halogen Bond. J. Wiley & Co., New York, 1973.

Received November 8, 1973

Stereochemistry of Propylene Polymerization

A. ZAMBELLI

Istituto di Chimica delle Macromolecole del C. N. R.
Via A. Corti 12, Milano, Italy

C. TOSI

Centro Ricerche Bollate, Montedison S. p. A.
Via S. Pietro 50, Bollate, Italy

Table of Contents

1. Preface 32

2. Polymer Structure 32
   2.1. Isotactic Polypropylene 32
   2.2. Syndiotactic Polypropylene 35
   2.3. Atactic Polypropylene 35

3. Polymerization Catalysts 36

4. Chain Propagation and Initiation 38

5. Monomer Insertion 39

6. Stereochemistry of the Addition to the Double Bond 41

7. Steric Control 41

8. Coordination 46

9. Catalytic Complexes 47
   9.1. Composition 47
   9.2. Structure 49

10. Propagation Mechanism 50
    10.1. Syndiotactic Propagation. Isotactic Propagation 51
    10.2. Anomalous Propagation. Isolated Errors. Stereoblocks. Errors of Chemical Arrangement 54

11. Conclusions 56

12. References 58

1. Preface

A number of reviews dealing with the stereospecific polymerization of α-olefins in the presence of Ziegler-Natta catalysts, or of monometallic catalysts closely resembling the Ziegler-Natta ones, are available in the literature (1). Therefore, we do not propose to write another comprehensive review article, but rather to express our views on the mechanism of polymerization, with particular reference to its stereochemical aspects. Stress is placed on the parts of the polymerization mechanism we consider to be soundly based on experimental evidence; furthermore, particular emphasis is given to work that seems to us to be fundamental to an understanding of the stereochemistry. In consequence this paper may appear both arbitrary and unilateral, such defects being overburdened by the preponderance inevitably given to our own work and that of the school to which we belong. Despite (or even because of?) this, we hope our article may serve to stimulate discussion and to spur on new research on the still controversial aspects of α-olefin polymerization.

2. Polymer Structure

In the presence of Ziegler-Natta catalysts α-olefins may be polymerized to either isotactic (2) or atactic polymers. Propylene may also be polymerized to syndiotactic polymer (3). The simplicity of the monomer structure, the commercial interest of the polymer, and the possibility of obtaining a variety of structures by different polymerization pathways (4, 5) render the study of propylene polymers particularly attractive.

2.1. Isotactic Polypropylene

Propylene can give rise to two monomer units of opposite configuration (Fig. 1).

The ideal structure of isotactic polypropylene (Fig. 2a) is a head-to-tail succession of monomer units of the same configuration (2). Depending on the catalyst and the synthesis conditions, the polymerization products will approximate to this ideal structure to a greater or lesser extent. In general, a raw polymer consists of a fraction insoluble in boiling

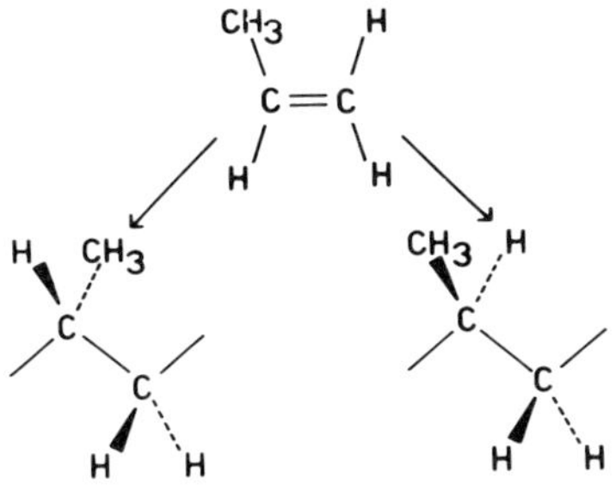

Stereochemistry of Propylene Polymerization

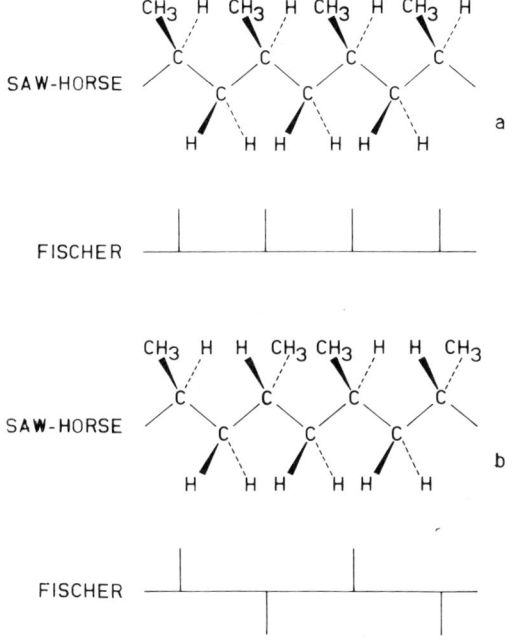

Fig. 1. Monomer units of propylene having opposite configurations (saw-horse drawing)

Fig. 2. Planar zigzag projection of the backbone carbon atoms of a) isotactic and b) syndiotactic polypropylene, both in the saw-horse and in the Fischer notation. It should be noted that the latter notation is not strictly a Fischer projection formula, because, by the Fischer convention, the carbon chain is always written vertically. In addition, only the methyl group positions are indicated here

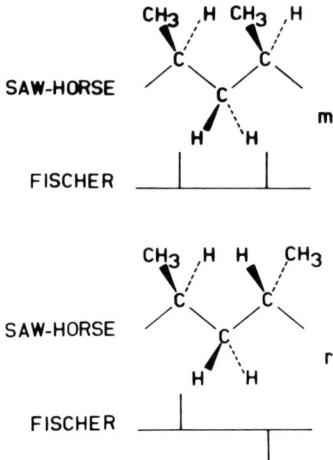

Fig. 3. Isotactic (m) and syndiotactic (r) diads in polypropylene: m (meso) and r (racemic) for the two diastereoisomeric diads refer to the notation of Frisch et al. (35)

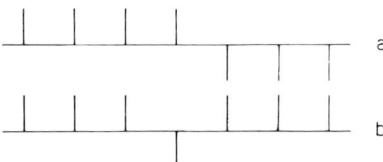

Fig. 4. Typical structural defects in isotactic polypropylene (simplified Fischer projection)

n-heptane, which melts above 170° C (6—10), and of more soluble fractions. The structure of the n-heptane insoluble fractions is closely akin to the ideal one (11, 12). The steric purity of isotactic polypropylene can be determined by several techniques. Since isotactic polypropylene is crystalline, and its degree of crystallinity depends *inter alia* on its steric purity, crystallinity measurements (*e.g.* through X-ray diffraction (13), and density (14) or melting point (8) determination) can be correlated with the steric purity. Direct measurements of steric regularity can be carried out by spectroscopic techniques (11, 12, 15—34). Among them, nmr analysis led to the conclusion (12) that, at least in many cases, the boiling *n*-heptane insoluble fractions of isotactic polypropylene consist to more than 98% of isotactic diads (Fig. 3). The main structural defects in these fractions (Fig. 4) are syndiotactic diads isolated between blocks of isotactic diads (*a* defects) (*36*) or, in lesser amount, pairs of syndiotactic diads (*b* defects) (*37*).

Instead, the more soluble fractions contain a larger number of syndiotactic diads. It is worth remarking that the distribution of the different diads within the chains is not Bernoullian: sequences of like diads are preferred over sequences of unlike diads, so that macromolecules of these fractions contain both isotactic stereoblocks and syndiotactic stereoblocks of variable length (*37*). The shorter the stereoblocks, the more soluble the macromolecules and the lower their melting points (*6*).

2.2. Syndiotactic Polypropylene

The ideal structure of this polymer (Fig. 2b) consists of a head-to-tail succession of monomer units of opposite configuration (*3*). The syndiotactic polymers of propylene obtained so far contain variable numbers of isotactic diads (Fig. 3); moreover, the diad distribution is not Bernoullian and in some cases the macromolecules contain, in addition to long syndiotactic stereoblocks, short isotactic stereoblocks (*37*).

2.3. Atactic Polypropylene

This polymer should ideally be formed by a head-to-tail succession of monomer units having randomly equal and opposite configurations. Although the term "atactic polypropylene" has been used in the literature in a very broad sense to indicate all noncrystallizable polymers, in fact no noncrystallizable polymer of propylene, whose structure has been studied so far, is strictly atactic (*12, 37*). Even in polymers in which the isotactic and syndiotactic diad content approaches 50%, the succession is not Bernoullian, since sequences of homo-diads are generally preferred over sequences of hetero-diads; hence these polymers also contain short isotactic stereoblocks near short syndiotactic stereoblocks[1].

---

[1] When we say: "These polymers contain isotactic (or syndiotactic) stereoblocks", we mean that equal diads are organized in sequences of average length longer than would be predicted by a Bernoullian distribution.

Atactic and syndiotactic polypropylenes always contain a certain proportion of units arranged head-to-head and tail-to-tail (38—41). In contrast, no such errors are generally present in the chemical arrangement of the monomer units of highly isotactic fractions (insoluble in boiling n-heptane). Both head-to-head and tail-to-tail units can be detected by ir spectroscopy [bands at 752 cm^{-1}, resulting from the rocking vibration of the group —CH$_2$—CH$_2$— bound on both sides to —CH(CH$_3$)— groups (42, 43), and at 1133 cm^{-1}, resulting from the vibration of the two methyls on adjacent methine groups] (38, 42, 43).

3. Polymerization Catalysts[2]

There are very many catalytic combinations capable of inducing Ziegler-Natta polymerization of α-olefins. Readers wishing to obtain a deeper insight into this subject should consult the review articles (1) as well as the original literature.

Some examples of catalyst systems that may be used to steer the propylene polymerization in the desired directions are shown in Table 1.

As can be seen from Table 1, the catalyst systems for propylene polymerization can be divided into classes, according to a number of characteristics. If we look at composition, we can distinguish a class of *monometallic* catalysts (which contain transition metals only) and a class of *bimetallic* catalysts (which also contain nontransition metals). According to physical criteria, it is possible to distinguish between *heterogeneous* catalysts (insoluble, at least in part, in the reaction medium) and *homogeneous* catalysts (soluble in the reaction medium). Finally, a distinction can be made between *isospecific, syndiospecific* and *nonspecific* catalysts, depending on the predominant stereospecificity. The empirical value of such classifications can hardly be overestimated, and they will be used, particularly the third one, in the present review. However, none of these classifications appears to be completely exhaustive from a speculative standpoint. In fact, polymers with an essentially similar fundamental structure can be obtained by means of both bimetallic and monometallic catalysts. Isotactic partially crystallizable polymers of α-olefins can be obtained with either heterogeneous or homogeneous catalysts (10, 45), as can also syndiotactic partially crystallizable polymers of propylene (3, 5, 46); the same holds for noncrystallizable polymers. A strict distinction depending on

---

[2] Supported catalysts are not considered here.

Table 1. Examples of catalyst systems for propylene polymerization

No.	Catalyst	Phase	Composition	Stereospecificity[a]	Polymerization temperature
1	$\alpha \text{TiCl}_3 — \text{Al}(C_2H_5)_2 J$ (10)	heterogeneous	bimetallic	isotactic	+70° C
2	δTiCl_3-amines (44)	heterogeneous	monometallic	isotactic	+70° C
3	$\text{Zr}(C_6H_5 — CH_2)_4$ (45)	homogeneous	monometallic	isotactic	+70° C
4	$\text{VCl}_4 — \text{Al}(C_2H_5)_2\text{Cl}$ (46)	homogeneous	bimetallic	syndiotactic	−78° C
5	$\text{VCl}_4 — \text{Al}(C_2H_5)_3$ (47)	homogeneous	bimetallic	non specific	−78° C

[a] By "stereospecificity" we mean the prevailing structure in the more abundant crystallizable fractions in the polymer. This structure might even not coincide with the type of propagation that predominates in the whole polymer (see Sections 4 and 7).

stereospecificity can hardly be made, since between the polymers with the highest isotactic stereoregularity and the polymers with the highest syndiotactic stereoregularity a nearly continuous range of products of intermediate structure can be prepared with suitable catalysts and by fractionation. Moreover, as mentioned above, many propylene polymers contain iso- and syndiotactic stereoblocks (*37*). It is thus impossible to make merely qualitative distinctions; the polymers can be better described by giving, for instance, the diad content and the average length of the iso- and syndiotactic stereoblocks. If we wish to classify the catalysts more rigorously according to stereospecificity, the only course is to assign to every catalyst a suitable index that accounts for the diad composition and "blockness". For example, starting from the examination of the polymerization products, one might deduce a compositional index (*e.g.* the fraction of syndiotactic diads in the polymer) between 0 and 1, and a "blockness" index, represented *e.g.* by the average length of stereoblocks, or by the product of the reactivity ratios $r_1 r_2$, if polymerizations of intermediate stereospecificity are regarded as copolymerizations of the two pseudomonomers "isotactic diad" and "syndiotactic diad"[3]. Since even noncrystallizable polymers do not seem to be strictly "atactic", the range of $r_1 r_2$ would be $1 < r_1 r_2 < \infty$.

The stereospecific behavior of a given catalyst depends on reaction conditions, particularly temperature. Therefore, the reaction conditions should be specified when trying to classify catalysts according to their stereospecificity.

4. Chain Propagation and Initiation

When chain growth is terminated by O-tritiated alcohols, at least some macromolecules contain terminal tritium atoms (*50*). Sometimes (*51*) ^{3}H is found in the polymers only if tritiated Brønsted acids stronger than methanol are used for the termination[4]. It can therefore

---

[3] The "product of reactivity ratios" $r_1 r_2$ can be related to other quantities for which more direct physical evidence is available, *e.g.* the parameter introduced by Chûjô (*48*) to describe deviations from randomness in terms of the number and weight averages of isotactic (or syndiotactic) diads. The analogy with the true copolymerization is remarkable: here a similar parameter, the "index of sequential homogeneity" (*49*), characterizes the distribution of sequence length of the two monomers.

[4] Especially for systems like 4 of Table 1.

be concluded that the chain propagation reaction takes place via monomer insertion into carbon–metal bonds that are hydrolyzed by the Brønsted acid:

propagation $M-(C_3H_6)_n-y + C_3H_6 \longrightarrow M-(C_3H_6)_{n+1}-y$

hydrolysis $M-(C_3H_6)_n-y + {}^3Hx \longrightarrow Mx + {}^3H-(C_3H_6)_n-y$

where M = metal atom; y = chain terminal; ${}^3Hx$ = tritiated Brønsted acid.

In the polymers prepared with catalytic systems containing organometallic compounds of nontransition metals (*e.g.* AlR_3, where R = alkyl, aryl or aralkyl radical) it has been established that $y = R$, at least for some macromolecules (52). In this case, initiation clearly consists of the insertion of the first propylene molecule into an M—R bond. In catalysts prepared in the absence of organometallic compounds an M—R bond is probably formed at first (it has not been ascertained through which reactions, although many examples of reactions forming metal–carbon bonds like the one proposed here have been reported in the literature) (53—57).

5. Monomer Insertion

The molecular weight of a polymer may be influenced by the presence of chain-transfer or termination reactions (58). One of these transfers leads to the cleavage of the M—P bond (P = growing macromolecule) with the formation of a dead macromolecule with terminal unsaturation. In isotactic polymers prepared at high temperature in the presence of the catalyst system $TiCl_3-Al(C_2H_5)_2Cl$, Longi *et al.* (59) detected the presence of terminal vinylidene unsaturation:

$$CH_2 = \overset{\overset{\displaystyle CH_3}{|}}{C} - (C_3H_6)_n - y$$

This fact was explained by assuming that chain propagation occurs via *primary insertion*, i.e. insertion into a metal–primary carbon bond, with formation of a new primary carbon–metal bond:

Primary insertion
(in the propagation)

$$M-CH_2-CH(CH_3)-(C_3H_6)_{n-y} + C_3H_6 \longrightarrow$$
$$\longrightarrow M-CH_2-CH(CH_3)-(C_3H_6)_{n+1}-y.$$

Takegami and Suzuki (60—62) studied insertion in the initiation reaction for some catalyst systems. They found primary insertion for the catalytic systems $TiCl_3$—$Al(C_2H_5)_3$ (I) (61) and VCl_4—$Al(C_2H_5)_3$ (II) (62), and secondary insertion for the catalytic systems VCl_4—$Al(C_2H_5)_2Cl$ (III) and VCl_4—$Al(C_2H_5)_2Cl$-anisole (IV) (60, 62):

Secondary insertion
(in the initiation)

$$M-C_2H_5 + C_3H_6 \longrightarrow M-CH(CH_3)-CH_2-C_2H_5$$

These authors also pointed out that system (I) gives rise to predominantly isotactic polymers, system (II) to noncrystallizable polymers, and systems (III) and (IV) to predominantly syndiotactic polymers. To provide an independent check of Takegami and Suzuki's conclusions, we have studied (41) the problem of the arrangement inversions in the predominantly syndiotactic polymers and in ethylene-propylene copolymers prepared in the presence of catalysts (III) and (IV). We deduced that chain propagation occurs mainly with secondary insertion in the presence of syndiospecific catalysts, and that the type of insertion is determined by steric factors. Thus while secondary insertion of a propylene unit is more frequent when the last unit of the growing chain is propylene, primary insertion is more frequent when the last unit is ethylene[5]. Summarizing these observations, we thought it justified (41) to assert the relationships shown in Table 2 between the last unit of the growing chain and insertion of the next propylene unit.

---

[5] If the rules that govern the type of insertion during syndiotactic chain propagation were to hold for initiation too, Takegami and Suzuki would also have found primary insertion in the case of syndiospecific catalysts (III) and (IV). In this case, in fact, a propylene unit adds in the initiation to a C_2H_5 group, the bulkiness of which is certainly not greater than that of a macromolecule ending in an ethylene unit.

Table 2. Mechanism of propylene insertion in the presence of syndiospecific catalytic systems

Last unit of the growing chain	Catalytic M—C bond		Newly formed catalytic M—C bond
Propylene	Secondary	$+ C_3H_6$	Secondary
	Primary	$+ C_3H_6$	Secondary
Ethylene	Primary	$+ C_3H_6$	Primary

It must be realized that the experimental data support only the first and third statements of the Table but not at all the second. What is certain is only that in this particular case secondary insertion on a secondary propylene is favored over primary insertion on a primary propylene. We stress this point, since we think it important, for a possible stereochemical reaction mechanism, to establish that isotactic propagation seems to occur with primary insertion and syndiotactic propagation with secondary insertion.

6. Stereochemistry of the Addition to the Double Bond

The addition of M—P (M = metal atom, P = polymer chain) to the double bond of a 1,2-disubstituted olefin CHA=CHB can give rise to two diastereoisomer monomeric units (Fig. 5). The structure of the monomer (*cis* or *trans*) and that of the monomeric units in the polymer (*erythro* or *threo*) may be correlated through the mechanism of addition to the double bond, as shown in Fig. 5.

Natta *et al.* (63), Miyazawa *et al.* (64) and Zambelli *et al.* (65) determined the structure of isotactic polymers of *cis* and *trans* 1-$d_1$propylene and of syndiotactic copolymers of d_6-propylene with *cis* and *trans* 1-d_1-propylene. In so doing, they succeeded in establishing that *cis* addition is common to both isotactic and syndiotactic propagation.

7. Steric Control

Soon after the discovery of isotactic polypropylene, the problem of the origin of the stereospecific trend of α-olefin polymerization has

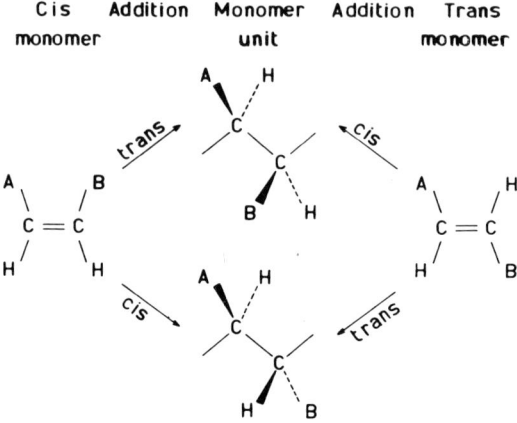

Fig. 5. Addition to the double bond. The attack of CHA is supposed to occur from the upper side of the paper. Should the attack of CHA occur from the lower side, the antipode monomer units would be formed

stimulated the interest of many authors. On the basis of rather indirect experimental evidence — if any — three main hypotheses were put forward in an attempt to explain the isospecific polymerization of propylene.

An isotactic macromolecule is formed through a series of successive asymmetric additions of the same sign on the monomer double bond:

$$M-[(d)CH_2CH(CH_3)]_n-P + C_3H_6 \longrightarrow M-[(d)CH_2CH(CH_3)]_{n+1}-P$$

where M = metal atom, P = segment of macromolecule, n = arbitrary number, (d) = configuration of the monomer units. A necessary (but not necessarily sufficient) condition for the occurrence of a series of transformations:

n prochiral monomers $\longrightarrow$ n units of chirality d

instead of a random series:

n prochiral monomers $\longrightarrow$ $\frac{1}{2}n$ units of chirality d
$\qquad\qquad\qquad\qquad\quad + \frac{1}{2}n$ units of chirality l

is that at least one of the reacting species must be chiral. The last unit of the growing chain contains a trisubstituted *d* or *l* carbon atom. This asymmetry alone could in theory be reason (*66*) for the addition

$$M-[(d)CH_2CH(CH_3)]_1-P + C_3H_6 \longrightarrow M-[(d)CH_2CH(CH_3)]_2-P$$

being favored over the addition

$$M-[(d)CH_2CH(CH_3)]-P + C_3H_6 \longrightarrow$$
$$\longrightarrow M-[(l)CH_2CH(CH_3)]-[(d)CH_2CH(CH_3)]-P$$

(*first hypothesis*).

On the other hand, the most favored conformation of isotactic polypropylene is a ternary helix (*67*); therefore, the addition

$$M-[(d)CH_2CH(CH_3)]_n-P + C_3H_6 \longrightarrow M-[(d)CH_2CH(CH_3)]_{n+1}-P$$

could be favored (*68*) over the diastereoisomeric addition, which would lead to $M-[(l)CH_2CH(CH_3)]-[(d)CH_2CH(CH_3)]_n-P$, according to the particular helicity (clockwise or anticlockwise) of the chain segment $[(d)CH_2CH(CH_3)]_n$ contiguous to the growing chain end (*second hypothesis*).

These two hypotheses are not identical, since the first attributes isospecific steric control to the 1—3 asymmetric induction from the last (and only the last) unit of the growing chain, while the second attributes steric control to the asymmetric induction of a helicoidal chain segment. Both could cause steric control, but it cannot be maintained that they actually do.

A *third hypothesis* concerning steric control has therefore been put forward (*69*). The M—P bond is part of an organometallic complex of unknown nature (either in solution or localized at the surface of insoluble catalysts). This complex could be asymmetric, due *e.g.* to a particular spatial arrangement of the ligands of M (*70*). Because of the chirality of M, the addition

$$(+)M-P + C_3H_6 \longrightarrow (+)M-[(d)CH_2CH(CH_3)]-P,$$

where (+) denotes the configuration of the catalytic complex, could be favored over the diastereoisomeric addition leading to (+)M—[(*l*)CH$_2$CH(CH$_3$)]—P, whatever the configuration of the last unit of the growing chain and the helicity of P. For the sake of clarity, it is worth remarking that the foregoing discussion refers to what happens in a single catalytic complex. Since, on the whole, the catalytic systems considered would be a racemic mixture of enantiomeric catalytic complexes, in addition to the described phenomena common to one half of the catalytic complexes, the mirror image phenomena should take place in the other half; these too would lead to isotactic propagation.

Syndiotactic polymerization also occurs through a series of asymmetric additions. In this series, however, additions of opposite configuration follow one another. The origin of the steric control could be that the addition

$$M—[(d)CH(CH_3)CH_2]—P + C_3H_6 \longrightarrow$$
$$\longrightarrow M—[(l)CH(CH_3)CH_2]—[(d)CH(CH_3)CH_2]—P$$

is favored over the diastereoisomeric addition, because of 1–3 induction from the last unit of the chain.

Alternatively, one could assume that M is chiral but its configuration inverts whenever a monomer molecule is added. If this were the case, the configuration of M could be the origin of the particular sequence of asymmetric additions which gives rise to syndiotactic propagation:

$$(+)M—P + C_3H_6 \longrightarrow (-)M—(d)(C_3H_6)—P$$
$$\{(+)M—P + C_3H_6 \longrightarrow (-)M—(l)(C_3H_6)—P\}$$
$$(-)M—(d)(C_3H_6)—P + C_3H_6 \longrightarrow (+)M—[(l)(C_3H_6)]—[(d)(C_3H_6)]—P$$
$$\{(-)M—(d)(C_3H_6)—C_3H_6 \longrightarrow (+)M—[(d)(C_3H_6)]—[(d)(C_3H_6)]—P\}$$

where the two additions within brackets should be less favored than the other two.

It seems to us that in this case chain helicity cannot be proposed as a cause of steric control, because the preferred conformation of syndiotactic polypropylene in solution is the planar zigzag one (*65, 71*).

In our opinion, the most direct experimental approach enabling a choice to be made between the previous hypotheses has been made by studying (via ^{13}C nmr) the structure of ethylene-propylene copolymers prepared with isospecific and syndiospecific catalysts (37, 72, 73). If the steric control of the isospecific polymerization were due to the asymmetry of the last unit of the chain, the addition of a propylene molecule to a growing chain ending in an achiral ethylene unit would not be stereospecific. Moreover, the presence of an ethylene unit at the end position of the growing chain would introduce a rotational freedom. This freedom should also eliminate any stereospecific influence of the helicity of the isotactic segment immediately preceding the ethylene unit on the addition of the next propylene molecule. Then, again in this case, the addition of a propylene unit should not be stereospecific. Therefore, among the possible causes of steric control, the only one which is not eliminated after addition of an ethylene molecule is the possible chirality of M. In fact, it has been observed that addition is isospecific, even in the step following the addition of an ethylene molecule (37, 73). Within the limits of confidence of the analytical method employed, the specificity of the reaction

$$M-(C_2H_4)-[(d)(C_3H_6)]-P + C_3H_6 \longrightarrow$$
$$\longrightarrow M-[(d)(C_3H_6)]-(C_2H_4)-[(d)(C_3H_6)]-P$$

with respect to the reaction

$$M-(C_2H_4)-[(d)(C_3H_6)]-P + C_3H_6 \longrightarrow$$
$$\longrightarrow M-[(l)(C_3H_6)]-(C_2H_4)-[(d)(C_3H_6)]-P$$

is complete. It can thus be concluded that, in isospecific polymerization, M is chiral and the steric control is due to the chirality of M.

In the case of syndiospecific polymerization the situation is more complex, since, as discussed in Section 5, propylene units which add to a growing chain ending with ethylene units preferentially give rise to primary (instead of secondary) addition (41). This fact alone would indicate that the last unit of the growing chain exerts a remarkable steric influence, which is not confined to the steric order but also involves the chemical arrangement.

Anyhow, if we assume that the addition of propylene is not stereospecific when an ethylene unit occupies the last place in the growing chain and take both types of insertion into account, the following pairs of diastereoisomeric sequences should be formed:

$$\begin{cases} \sim (d)[CH_2CH(CH_3)]\text{—}CH_2CH_2\text{—}(d)[CH(CH_3)CH_2] \sim & \text{(1a)} \\ \sim (d)[CH_2CH(CH_3)]\text{—}CH_2CH_2\text{—}(l)[CH(CH_3)CH_2] \sim & \text{(1b)} \end{cases}$$

$$\begin{cases} \sim (d)[CH_2CH(CH_3)]\text{—}CH_2CH_2\text{—}(d)[CH_2CH(CH_3)] \sim & \text{(2a)} \\ \sim (d)[CH_2CH(CH_3)]\text{—}CH_2CH_2\text{—}(l)[CH_2CH(CH_3)] \sim & \text{(2b)} \end{cases}$$

in equal amounts two by two (1a = 1b and 2a = 2b). Two pairs of diastereoisomeric sequences are actually observed, but the amounts of 1a and 2a seem to be slightly different from those of 1b and 2b (72).

The most plausible interpretation of this behavior might be that, in syndiospecific polymerization, steric control is essentially due to asymmetric 1–3 induction from the last unit of the chain. The differences found in the amounts of 1a and 1b, and 2a and 2b, could also indicate the presence of partial steric control due to M (which would then be asymmetric). Such control would be convergent, or more probably divergent, with respect to the more important control by the last unit of the growing chain; furthermore, it is quantitatively negligible compared to the latter, so that it would become detectable only when the last unit is ethylene.

In conclusion, the isospecific steric control should be due to the chirality of M, whereas the syndiospecific steric control should be due to the chirality of the last unit of the growing chain end. This conclusion also agrees with the statistics of sequence distribution observed in ethylene-propylene copolymers (74).

8. Coordination

The polymerization mechanism of α-olefins is often defined as anionic-coordinated. The term "anionic", if intended to denote an asymmetric electronic distribution on the M—P bond with a higher electronic density on P than on M, seems justified (see hydrolysis with tritiated Brønsted acids, Section 4). The term "coordinated" is used to stress the assumption that, like other reactions catalyzed by complexes of transition metals, propylene should coordinate at the transition metal before insertion into the M—P bond. Polymerization would then take place in two successive steps:

coordination $\quad\quad C_3H_6 + M\text{—}P \longrightarrow C_3H_6 \cdot (M\text{—}P)$

insertion $\quad\quad\quad C_3H_6 \cdot (M\text{—}P) \longrightarrow M\text{—}(C_3H_6)\text{—}P$

As remarked by Boor (*1*), direct experimental evidence supporting this view is scanty but not, in our opinion, insignificant.

Catalytic systems, isospecific with respect to propylene, have also been used to polymerize vinyl aromatic monomers; in this case too, crystalline isotactic polymers are obtained, so that it is reasonable to assume that the same mechanism governs the two polymerizations. Natta, Danusso and Sianesi (*75—78*) studied the polymerization and the binary copolymerization of a series of vinyl-aromatic compounds characterized by constant bulkiness and variable electronic density on the double bond. They observed that, for two given monomers A and B, $r_1 = V_A/V_B = 1/r_2$, where r_1, r_2 are the reactivity ratios and V_A, V_B are the homopolymerization rates of A and B under equal conditions. Furthermore, the V's of the studied monomers obeyed the Hammett equation (*79*) with negative ϱ. Therefore, the propagation rate would be determined by a step of electrophilic attack on the double bond. Most probably the electrophilic agent is a transition metal atom at which the monomer coordinates, and the rate-determining step is this coordination.

For syndiotactic polymerization the only evidence for coordination is the peculiar dependence of the polymerization rate on the monomer concentration. The form of such dependence seems to indicate that there are two successive reactions, both rate-determining; these reactions could be coordination and insertion (*80*).

9. Catalytic Complexes

9.1. Composition

By "catalytic complexes" we mean the chain-propagating species, superficial or in solution, depending on whether the catalyst is heterogeneous or homogeneous. They differ from the catalyst (or catalytic system), even when it is homogeneous or prepared from a single, well-defined chemical compound. For instance, Giannini et al. (45) have shown that, even when α-olefin polymerization is catalyzed by tetrabenzyltitanium, the active species is not tetrabenzyltitanium itself, but a transformation product of unknown structure. The number of active species, with the possible exception of some supported catalysts, is negligible as compared with the molar amount of the catalyst components. Whether because of this, or the low thermal stability of many and perhaps all catalytic complexes, attempts to isolate the true active species have not so far been successful. Nevertheless, it seems well established that, in the active species M—P, M is a transition metal. For the many arguments supporting this statement, the reader is referred to an excellent review article (1). We will simply quote the argument that seems to us the most cogent, namely the fact that monometallic catalyst systems exist at all (81). Several speculations have also been made regarding the ligands (in addition to P) of the catalytic complexes. In the monometallic, heterogeneous isospecific catalyst systems, the ligands of the superficial active species cannot be other than those present from the start on the crystal surface. But is this a general property of all heterogeneous catalyst systems, or is it a particular one, in the sense that in heterogeneous bimetallic systems there can also be isospecific active species, where some of the ligands stem from the formation of superficial bimetallic complexes (e.g. via chlorine bridges) with the organometallic compound of the nontransition metal? This problem was reviewed by Boor (1) and we refer the reader to him for the arguments for and against the two hypotheses; in our opinion, though, the question is still open.

From a logical point of view, it is quite attractive to assume a uniqueness of composition of isospecific superficial catalytic complexes (strictly "monometallic" hypothesis), since isospecific superficial monometallic active species certainly exist. However, it is rather difficult to explain by this hypothesis alone the considerable differences in overall catalytic activity observed in different systems (82). Moreover, many arguments reported in the same review (1) and emerging from a comparison of the catalytic systems 4 and 5 of Table 1, lead one to assume with a fair degree of confidence that, at least in homogeneous bimetallic catalyst systems, bimetallic catalytic complexes are present too. Therefore it seems as much logically attractive to assume that catalytic complexes can also occur on the surface of heterogeneous catalysts, where transition and non-transition metal atoms share some ligands.

We believe that no general statement can be made on the oxidation state of the transition metals in catalytic complexes that are active in the α-olefin polymerization; this state could differ from one system to another, and the literature data are not in agreement.

9.2. Structure

Observations by electron microscopy have shown that, when polymerization is carried out on well-formed crystals of α-TiCl$_3$, polymer formation occurs only on the lateral faces and the edges of crystals, but not on the basal faces (*83*). The layer structure of α-TiCl$_3$ appears to be such that on the lateral faces and the edges there are "exposed" titanium atoms, i.e. atoms bearing coordination vacancies, while the basal faces consist of planes of chlorine atoms overlying lattice planes occupied by nonexposed titanium atoms, i.e. atoms located at the center of a coordination polyhedron whose apices are occupied by six Cl's, each one shared by two titanium atoms (Fig. 6). The titanium

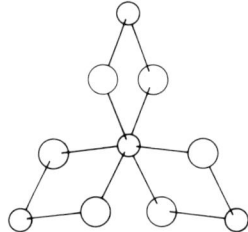

Fig. 6. The coordination polyhedron in α-TiCl$_3$ (projection along the trigonal axis of the layer). Large circles represent Cl ligands and small circles represent Ti atoms

atom of the active species M—P could then reasonably be expected to be exposed. After consideration of monometallic isospecific complexes, Cossee and Arlman (*84—87*) proposed a model of isospecific superficial catalytic complexes in which the titanium atom is octahedral, the ligands being four chlorine atoms, the polymeric radical P, and a co-ordinated monomer molecule. The asymmetry of the titanium atom required by the type of steric control may be found in the nonequivalence of the Cl atoms, partly lying on the side surface of the layer and partly not (Fig. 7). Cossee's model is particularly appealing, despite a few inconveniences which some authors have tried to eliminate (*88*), because of its mechanistic implications and because it meets the requirement that complexes on which the isospecific propagation takes place must be asymmetric.

As regards the homogeneous vanadium-based catalytic complexes on which propagation is predominantly syndiospecific, similar coordi-

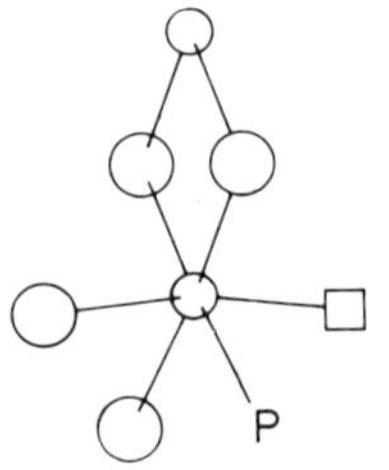

Fig. 7. Supposed structure of the active center in a Ziegler-Natta catalyst. P is the growing polymer chain and the square is the vacant octahedral position

nation polyhedra have been proposed. The chief difference would be that, though the V atom could still be asymmetric, the forces of steric control due to this possible asymmetry would be insignificant because of the steric equivalence of the Cl's or, more generally, of the ligands. We point out again that, as discussed in Section 7, syndiotactic propagation requires the steric control to be attributed to the asymmetry of the last unit of the chain, but it does not mean that the catalytic complexes must be achiral[6].

10. Propagation Mechanism

Contrary to what the title of this section seems to imply, we do not intend to review the propagation mechanisms proposed by several authors (70, 84, 89) that have stimulated so much research, nor do we even propose our own mechanism. We merely try to assemble the experimental facts previously described, each time selecting hypotheses that, if not original, have at least the merit of simplicity. We shall also neglect what is probably the most important problem, viz. the capability of carbon-transition metal bonds to add to the olefinic double bond. We merely recall that Cossee and Arlman gave a logical picture of the heterogeneous Ziegler-Natta catalysis in the polymerization of α-olefins in terms of quantum chemistry and of crystal chemistry (84–86).

---

[6] Analogously, isotactic propagation requires that steric control be assigned to the chirality of M, but does not impose the condition that the last unit of the chain be achiral. It will be readily appreciated that such a requirement would render isotactic propagation a contradiction in terms.

10.1. Syndiotactic Propagation. Isotactic Propagation

At this stage it is appropriate to list the peculiar features of isotactic and syndiotactic propagation (Table 3).

Table 3. Peculiar features of isotactic and syndiotactic polymerization

	Isotactic propagation	Syndiotactic propagation
Addition to the double bond	cis	cis
Monomer insertion	Primary	Secondary
Chiral center of steric control	M	Last unit of the growing chain

As remarked by Cossee (84, 86, 90), the *cis* type of addition to the double bond joins the olefin polymerization to the many micromolecular reactions of "*cis* ligand migration"[7] catalyzed by complexes of transition metals (oxosynthesis, hydrogenation, etc.), and suggests an activated four-center complex (Fig. 8).

Both primary and secondary insertions are encountered in the *cis* ligand migration reactions. One is favored over the other to an extent that depends on electronic factors (the particular transition metal, the nature of ligands) and on reaction conditions (temperature, solvent, etc.). Where electronic factors do not wholly preclude one of the possible insertions, steric factors can also play a role.

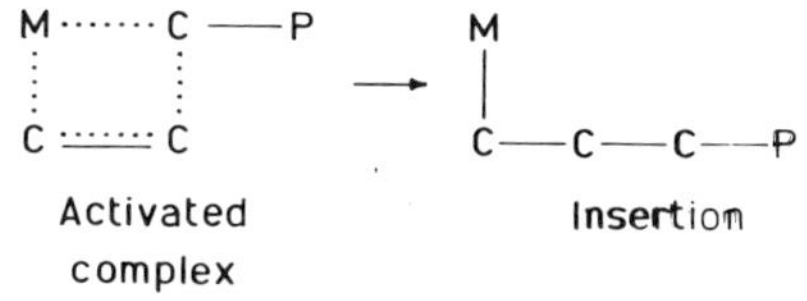

```
M ······ C ——— P              M
:         :          →        |
C ······ C                    C ——— C ——— C ——— P
  Activated                    Insertion
  complex
```

Fig. 8. The four-center activated complex through which the addition of M—P to the monomer double bond should occur

---

[7] In the olefin polymerization, however, the addition could be a *cis* ligand insertion. We shall use the two terms indifferently, since we have no reason to prefer one mechanism to another, and the resulting uncertainty does not interfere with our discussion.

For instance, in the case of hydroformylation of α-olefins in the presence of simple carbonyl catalysts, both branched and linear aldehydes are formed, whereas with some substituted carbonyl catalysts essentially linear aldehydes are obtained (91, 92). Branched aldehydes are believed to be formed via secondary insertion of the olefin into a σ H—M bond, and linear aldehydes via primary insertion:

$$\begin{array}{c} R \\ | \\ H_2C=CH \\ \downarrow \\ H-M\cdot CO \end{array} \longrightarrow \begin{array}{c} CH_3-CH-R \\ | \\ M\cdot CO \end{array} \longrightarrow \begin{array}{c} CH_3-CH-R \\ | \\ M\cdot C=O \end{array} \xrightarrow{H_2} \begin{array}{c} CH_3-CH-R \\ | \\ CHO \end{array}$$

$$\begin{array}{c} R-CH=CH_2 \\ \downarrow \\ H-M\cdot CO \end{array} \longrightarrow \begin{array}{c} R-CH_2-CH_2 \\ | \\ M\cdot CO \end{array} \longrightarrow \begin{array}{c} R-CH_2-CH_2 \\ | \\ M\cdot C=O \end{array} \xrightarrow{H_2} \begin{array}{c} R\cdot CH_2-CH_2 \\ | \\ CHO \end{array}$$

It is therefore not surprising that two types of insertion are also found in polymerization, the nature of M and its ligands favoring one insertion rather than another. The secondary insertion with formation of an activated four-center complex via attack in the direction of the roomier side with respect to the substituted carbon of the last unit of the growing chain explains how syndiotactic propagation occurs through steric interactions of the monomer with the last unit (62). The activated *trans* complex (Fig. 9a) should give rise to weaker non-bonded interactions than the *cis* complex (Fig. 9b), provided the steric environments above and below M be equal or only slightly different (as would happen *e.g.* if syndiospecific catalytic complexes were to have the structure proposed in Section 9). We note that the (stabler) activated

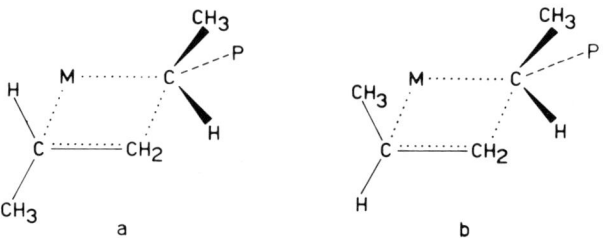

Fig. 9. *Trans* (a) and *cis* (b) activated complexes for typical secondary insertion

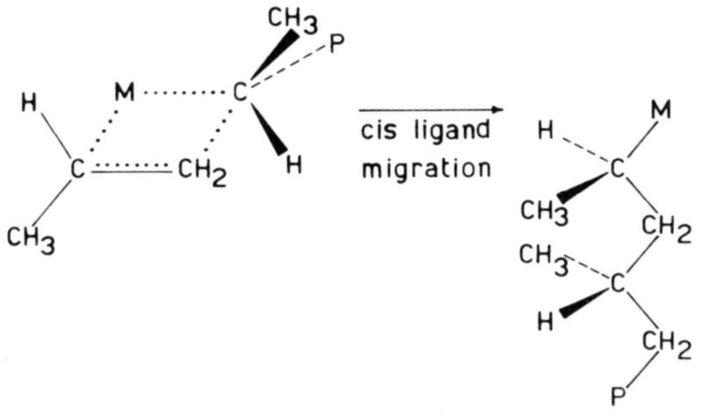

Fig. 10. Syndiotactic polypropylene chain resulting from the *trans* activated complex via *cis* ligand migration

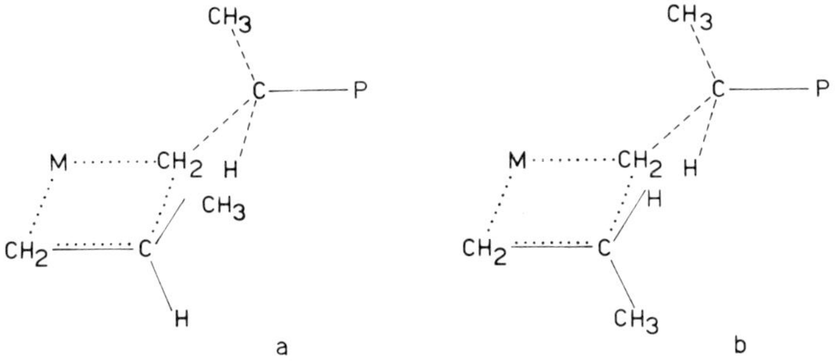

Fig. 11. Supposed four-center activated complexes for a typical primary insertion of a propylene molecule

trans complex leads to syndiotactic propagation via *cis* ligand migration (Fig. 10), whereas the (less stable) *cis* complex would lead to isotactic propagation. Conversely, the formation of activated four-center complexes via primary insertion leads one to minimize the differences of nonbonded interactions between the monomer substituent and the substituent of the last unit in the two diastereoisomers shown in Fig. 11, whatever the monomer presentation, because of the nonrigid position (rotation about the C—C bond) of the tertiary carbon atom of the last unit of the chain, which is in this case outside the four-membered ring.

Under these conditions, the possible presence of a steric control depends only on the possible difference of the steric environment above and below M. If the bulkiness above and below M is appreciably different, and the monomer approaches the reactive M—P bond from a constant direction, as predicted *e.g.* by Cossee's model of isospecific complexes (*84*), primary insertion will lead to prevailingly isotactic propagation, whatever the last unit of the growing chain (even if it is an ethylenic unit).

It is now apparent that, according to the foregoing scheme, isotactic[8] or syndiotactic propagation could in principle take place on the same catalyst complex, depending on whether insertion is primary or secondary.

10.2. Anomalous Propagation. Isolated Errors. Stereoblocks. Errors of Chemical Arrangement

As mentioned in Section 2, the sterically purest isotactic fractions insoluble in boiling *n*-heptane contain traces of isolated syndiotactic diads (Fig. 4a) and of adjacent syndiotactic diads (Fig. 4b). At least in some cases, the fraction of adjacent syndiotactic diads is not so low as to be reconciled with the assumption that their presence is simply due to random successive repetition of errors in the steric control. Hence, it is probable that different mechanisms are involved in the two types of steric defects. On the other hand, if we take the mechanism of steric control into account, we can see that the formation of pairs of adjacent syndiotactic diads could be easily explained as a single violation of the steric control in the addition of a monomer molecule. The presence of isolated syndiotactic diads entails not a single violation of the steric control, but a complete inversion of its sign. If the isotactic segment preceding the syndiotactic diad had formed through a series of *d* additions, the next segment would form through a series of *l* additions (Fig. 12). A hypothesis capable of reconciling this event with the origin of the isospecific steric control consists in assuming that the configuration of the catalytic complexes can sometimes invert.

All other polymers (crystallizable polymers with mainly syndiotactic structure, noncrystallizable polymers, and crystallizable fractions soluble

---

[8] Isotactic propagation should be more or less free from defects, depending on how much the bulkiness above M differs from the bulkiness below M. The stereospecific driving forces occurring in the insertion according to the present mechanism do not seem quantitatively adequate when highly isotactic polymers (see Section 2.1) are obtained by heterogeneous catalyst systems; in these cases an even more cogent steric control could formerly occur in the coordination step (*88*). This view would agree (see Section 8) with the arguments supporting a rate determining coordination of the monomer.

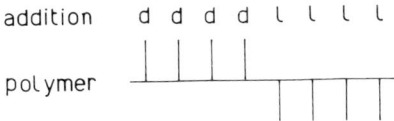

Fig. 12. Segment of a polypropylene chain formed by a series of d additions followed by a series of l additions

in boiling n-heptane with mainly isotactic structure) simultaneously contain different numbers of isotactic and syndiotactic stereoblocks (see Sections 2 and 3). In these polymers there are also errors of chemical arrangement.

As explained in the foregoing section, there exists a relationship between the type of propagation (isotactic or syndiotactic) and the manner of monomer insertion (primary or secondary). The presence of head-to-head and tail-to-tail units suggests, therefore, that stereoblocks could be formed simply because, after a certain number of consecutive primary insertions (disordered, or more or less isotactic block), a number of secondary insertions (syndiotactic block) occur. This explanation has the advantage that it does not imply any change in the structure of the catalytic complex; at present, however, it is nothing more than a working hypothesis. It seems to us that the mechanistic elements reviewed above render it particularly suggestive, also because they allow us to foresee that, if both types of insertion are possible on a catalytic complex (for instance, as in oxosynthesis reaction), these will not follow one another randomly, but insertions of the same type will tend to sequentiate in such a way as to permit formation of predominantly isotactic and syndiotactic stereoblocks separated by head-to-head or tail-to-tail units.

Let us imagine that primary insertion is initially favored on a given catalytic complex for electronic and/or steric reasons. Under these conditions, the steric interactions of monomer with the ligands of M are minimized in the activated complex, thus leading to a further primary insertion (Fig. 11a). However, after the random occurrence of a secondary insertion, the probability of a new secondary insertion increases because of the steric interactions that would otherwise take place between the secondary carbon of the last unit of the growing chain and the substituted carbon of the monomer: in the case of a primary insertion of the new monomer, these two carbons would occupy vicinal positions (Fig. 13).

A clue supporting this hypothesis might be the fact mentioned in Section 5 that, in the presence of syndiospecific catalysts based on

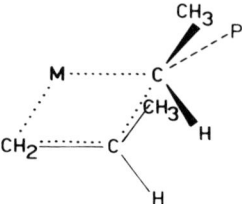

Fig. 13. Activated complex for a primary insertion on a secondary carbon-metal bond

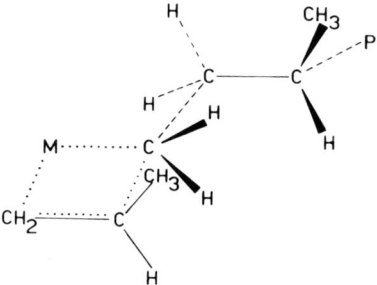

Fig. 14. Primary insertion of a propylene molecule on a growing chain ending in an ethylenic monomer unit

vanadium compounds, syndiotactic propagation takes place predominantly (or exclusively?) via secondary insertion of the monomer into secondary M—P bonds, while the monomer insertion becomes predominantly primary when the substituent of the last unit of the chain is removed by the prior entrance into the chain of an ethylenic unit (Fig. 14).

11. Conclusions

In conclusion, the experimental evidence reported in this review appears to support the following mechanistic hypotheses concerning the Ziegler-Natta polymerization of propylene:

1. Chain propagation occurs via monomer insertion into carbon–transition metal bonds M—P.

2. The addition of M—P to the monomer double bond occurs through an activated four-center complex.

3. The monomer orientation in the four-center complex can give rise to either primary or secondary insertion.

4. The type of insertion depends on a number of electronic and steric factors. Of the latter, the bulkiness of the M ligands tends to favor primary insertion, whereas the presence of a secondary last unit in the growing chain end favors secondary insertion of the next unit. Consequently, in the series of insertions leading to the formation of a macromolecule, the tendency toward a succession of identical insertions will be greater than toward a succession of unlike insertions, even in the limiting case of a 50%–50% distribution of primary and secondary insertions.

5. Secondary insertion into a secondary M—P bond makes possible the steric control of the last unit of the chain, and leads to syndiotactic propagation. Conversely, isotactic propagation is controlled by the asymmetry of the steric environment of the transition metal; it becomes possible when the primary insertion eliminates the steric control of the last unit of the growing chain end.

6. The structure of real macromolecules sometimes shows deviations from ideality (true isotacticity or syndiotacticity) because of the presence of errors in the steric control, the possibility of catalyst isomerization, and the nonconstancy of the type of insertion which leads to the presence of more or less isotactic and syndiotactic stereoblocks in the same macromolecule.

7. At least in some cases, noncrystallizable polymers of propylene are such, not because of a complete lack of stereospecificity of propagation, but because of frequent changes in the type of steric control during propagation.

8. The bulkiness above M is more likely to differ from that below M when catalytic complexes are on a crystal surface. Consequently, isospecific catalysts are more frequently heterogeneous and syndiospecific catalysts are more frequently homogeneous. This does not rule out that both types of stereospecific propagation can occur on both heterogeneous and homogeneous complexes.

12. References

1. Boor, J., Jr.: The Nature of the Active Site in the Ziegler-Type Catalyst, J. Polymer Sci., Part D **2**, 115 (1967). This article contains a very large number of references on earlier works. For more recent literature, see *e.g.*: Boor, J., Jr.: Ind. Eng. Chem. Prod. Res. Develop. **9**, 437 (1970). – Tsuruta, T.: J. Polymer Sci., Part D **6**, 179 (1972).
2. Natta, G., Corradini, P.: Atti Accad. Naz. Lincei, Memorie (8), **4**, 73 (1955). – Natta, G., Pino, P., Corradini, P., Danusso, F., Mantica, E., Mazzanti, G., Moraglio, G.: J. Am. Chem. Soc. **77**, 1708 (1955).
3. — Pasquon, I., Corradini, P., Peraldo, M., Pegoraro, M., Zambelli, A.: Rend. Accad. Naz. Lincei (8), **28**, 539 (1960).
4. Natta, G.: J. Polymer Sci. **16**, 143 (1955).
5. — Pasquon, I., Zambelli, A.: J. Am. Chem. Soc. **84**, 1488 (1962).
6. — Mazzanti, G., Crespi, G., Moraglio, G.: Chim. Ind. (Milano) **39**, 275 (1957).
7. Corradini, P.: Rend. Ist. Lombardo Sci. Lettere **91**, 889 (1957).
8. Natta, G.: Rend. Accad. Naz. Lincei (8), **24**, 246 (1958).
9. Danusso, F., Moraglio, G., Flores, E.: Rend. Accad. Naz. Lincei (8), **25**, 520 (1958).
10. Natta, G., Pasquon, I., Zambelli, A., Gatti, G.: J. Polymer Sci. **51**, 387 (1961).
11. Zambelli, A., Segre, A. L., Farina, M., Natta, G.: Makromol. Chem. **110**, 1 (1967).
12. — Zetta, L., Sacchi, C., Wolfsgruber, C.: Macromolecules **5**, 440 (1972).
13. See *e.g.*: Natta, G., Corradini, P., Cesari, M.: Rend. Accad. Naz. Lincei (8), **22**, 11 (1957).
14. Quynn, R. G., Riley, J. L., Young, D. A., Noether, H. D.: J. Appl. Polymer Sci. **2**, 166 (1959).
15. Luongo, J. P.: J. Appl. Polymer Sci. **3**, 302 (1960).
16. Koenig, J. L., Van Roggen, A.: J. Appl. Polymer Sci. **9**, 359 (1965).
17. Hughes, R. H.: J. Appl. Polymer Sci. **13**, 417 (1969).
18. Heatley, F., Zambelli, A.: Macromolecules **2**, 618 (1969).
19. Zambelli, A., Segre, A. L.: J. Polymer Sci., Part B, **6**, 473 (1968).
20. Stehling, F. C.: J. Polymer Sci., Part A, **2**, 1815 (1964).
21. Woodbrey, J. C.: J. Polymer. Sci., Part B, **2**, 315 (1964).
22. Kato, Y., Nishioka, A.: Bull. Chem. Soc. Japan **37**, 1622 (1964).
23. Woodbrey, J. C., Trementozzi, Q. A.: J. Polymer Sci., Part C **8**, 113 (1965).
24. Tincher, W. C.: Makromol. Chem. **85**, 34 (1965).
25. Ohnishi, S., Nukada, K.: J. Polymer Sci., Part B, **3**, 179, 1001 (1965).
26. Natta, G., Lombardi, E., Segre, A. L., Zambelli, A., Marinangeli, A.: Chim. Ind. (Milano) **47**, 378 (1965).
27. Boor, J., Jr., Youngman, E. A.: J. Polymer Sci., Part A-1, **4**, 1861 (1966).
28. Lombardi, E., Segre, A. L., Zambelli, A., Marinangeli, A., Natta, G.: J. Polymer Sci., Part C, **16**, 2539 (1967).
29. Kissin, Yu. V., Tsvetkova, V. I., Chirkov, N. M.: Europ. Polym. J. **8**, 529 (1972).
30. Satoh, S., Chûjô, R., Ozeki, T., Nagai, E.: J. Polymer Sci. **62**, S 101 (1962).
31. Heatley, F., Salovey, R., Bovey, F. A.: Macromolecules **2**, 619 (1969).
32. Ferguson, R. C.: Macromolecules **4**, 324 (1971).
33. Vasko, P. D., Koenig, J. L.: Macromolecules **3**, 597 (1970).
34. Johnson, L. F., Heatley, F., Bovey, F. A.: Macromolecules **3**, 175 (1970).
35. Frisch, H. L., Mallows, C. L., Bovey, F. A.: J. Chem. Phys. **45**, 1505 (1966).

36. Bovey, F. A.: NMR Basic Principles and Progress, Vol. 4, p. 1.Berlin - Heidelberg - New York: Springer 1971.
37. Zambelli, A.: NMR Basic Principles and Progress, Vol. 4, p. 101. Berlin - Heidelberg - New York: Springer 1971.
38. Van Schooten, J., Mostert, S.: Polymer **4**, 135 (1963).
39. Natta, G., Valvassori, A., Ciampelli, F., Mazzanti, G.: J. Polymer Sci., Part A, **3**, 1 (1965).
40. Tosi, C., Valvassori, A., Ciampelli, F.: European Polymer J. **5**, 575 (1969).
41. Zambelli, A., Tosi, C., Sacchi, C.: Macromolecules **5**, 649 (1972).
42. Ciampelli, F., Tosi, C.: Spectrochim. Acta **24** A, 2157 (1968).
43. Tosi, C., Zerbi, G.: Chim. Ind. (Milano) **55**, 334 (1973).
44. Boor, J., Jr., Youngman, E. A.: J. Polymer Sci., Part B, **2**, 265 (1964).
45. Giannini, U., Zucchini, U., Albizzati, E.: J. Polymer Sci., Part B, **8**, 405 (1970).
46. Zambelli, A., Natta, G., Pasquon, I.: J. Polymer Sci., Part C, **4**, 411 (1963).
47. — Natta, G., Pasquon, I., Signorini, R.: J. Polymer Sci., Part C, **16**, 2485 (1967).
48. Chûjô, R.: J. Macromol. Sci., Part B-2, **1**, 1 (1968).
49. Tosi, C., Catinella, G.: Makromol. Chem. **137**, 211 (1970).
50. Feldman, C. F., Perry, E.: J. Polymer Sci. **46**, 217 (1960).
51. Zambelli, A., Sacchi, C.: Makromol. Chem. (in press).
52. Natta, G., Pasquon, I.: Adv. Catal., Vol. 11, Ch. 11, p. 1. New York: Academic Press 1959.
53. Farrod, J. F., Chalk, A.: J. Am. Chem. Soc. **86**, 1776 (1964); **88**, 3491 (1966).
54. Green, M. L. H., Nagy, P. L. I.: J. Organomet. Chem. **1**, 58 (1963).
55. Sternberg, H. W., Wender, I.: Chem. Soc. (London) Spec. Publ. **13**, 35 (1959).
56. Manuel, T. A.: J. Org. Chem. **27**, 3941 (1962).
57. Cramer, R., Lindsey, R. V., Jr.: J. Am. Chem. Soc. **88**, 3534 (1966).
58. See *e.g.*: Natta, G., Pasquon, I., Giachetti, E., Pajaro, G.: Chim. Ind. (Milano) **40**, 267 (1958).
59. Longi, P., Mazzanti, G., Roggero, A., Lachi, M. P.: Makromol. Chem. **61**, 63 (1963).
60. Takegami, Y., Suzuki, T.: Bull. Chem. Soc. Japan **42**, 848 (1969).
61. — — Okazaki, T.: Bull. Chem. Soc. Japan **42**, 1060 (1969).
62. Suzuki, T., Takegami, Y.: Bull. Chem. Soc. Japan **43**, 1484 (1970).
63. Natta, G., Farina, M., Peraldo, M.: Chim. Ind. (Milano) **42**, 255 (1960).
64. Miyazawa, T., Ideguchi, T.: J. Polymer Sci., Part B, **1**, 389 (1963).
65. Zambelli, A., Giongo, M. G., Natta, G.: Makromol. Chem. **112**, 183 (1968).
66. Ohsumi, Y., Higashimura, I., Okamura, S.: J. Polymer Sci., Part A, **3**, 3729 (1965).
67. Natta, G., Corradini, P.: Nuovo Cimento, Suppl. (10), **15**, 40 (1960).
68. Coover, H. W., Jr.: J. Polymer Sci., Part C, **4**, 1511 (1963).
69. Natta, G.: J. Inorg. Nucl. Chem. **8**, 589 (1958).
70. Cossee, P.: Tetrahedron Letters **17**, 12 (1960).
71. Allegra, G., Ganis, P., Corradini, P.: Makromol. Chem. **61**, 225 (1963).
72. Zambelli, A., Gatti, G., Sacchi, C., Crain, W. O., Jr., Roberts, J. D.: Macromolecules **4**, 475 (1971).
73. Crain, W. O., Jr., Zambelli, A., Roberts, J. D.: Macromolecules **4**, 330 (1971).
74. Zambelli, A., Léty, A., Tosi, C., Pasquon, I.: Makromol. Chem. **115**, 73 (1968).
75. Natta, G., Danusso, F., Sianesi, D.: Makromol. Chem. **30**, 237 (1959).
76. Sianesi, D., Pajaro, G., Danusso, F.: Chim. Ind. (Milano) **41**, 1176 (1959).
77. Danusso, F.: Chim. Ind. (Milano) **44**, 611 (1962).
78. — Sianesi, D.: Chim. Ind. (Milano) **44**, 493 (1962).

79. Hammett, L. P.: Physical Organic Chemistry, p. 184. New York: McGraw-Hill 1940.
80. Zambelli, A., Pasquon, I., Signorini, R., Natta, G.: Makromol. Chem. **112**, 160 (1968).
81. See *e.g.*: Oita, K., Nevitt, T. D.: J. Polymer Sci. **43**, 585 (1960).
82. Zambelli, A., Pasquon, I., Marinangeli, A., Lanzi, G., Mognaschi, E. R.: Chim. Ind. (Milano) **46**, 1464 (1964).
83. Hargitay, B., Rodriguez, L., Miotta, M.: J. Polymer Sci. **35**, 559 (1959).
84a. Cossee, P.: J. Catalysis **3**, 80 (1964).
84b. Arlman, E. J.: J. Catalysis **3**, 89 (1964).
84c. — Cossee, P.: J. Catalysis **3**, 99 (1964).
85. — J. Catalysis **5**, 178 (1966).
86a. Cossee, P., Ros, P., Schachtschneider, J. H.: 4th International Congress on Catalysis, Moscow 1968, Preprint of the paper 14.
86b. — "The Stereochemistry of Macromolecules". Ed. A. D. Ketley, Marcel Dekker, New York 1967, Vol. I, Chapter 3, p. 145.
87. Armstrong, D. R., Perkins, P. G., Steward, J. J. P.: J. Chem. Soc. (Dalton Trans.), 1972 (1972).
88. Allegra, G.: Makromol. Chem. **145**, 235 (1971).
89. See Refs. 40, 52, 53, 63, 65–74, 77–82 of the Boor's review (*1*).
90. Cossee, P.: Rec. Trav. Chim. **85**, 1151 (1966).
91. Hughes, V. L., Kirshenbaum, I.: Ind. Eng. Chem. **49**, 1999 (1957).
92. Pruett, R. L., Smith, J. A.: J. Org. Chem. **34**, 327 (1969); — Craddock, J. H., Hershman, A., Paulik, F. E., Roth, J. F.: Ind. Eng. Chem. Prod. Res. Develop. **8**, 291 (1969).

Received October 16, 1973

Mercaptan-Containing Polymers

CHIEN-DA S. LEE

3M Company, 3M Center, 230-2S, St. Paul, Minnesota, 55101/USA

W. H. DALY

Department of Chemistry, Louisiana State University, Baton Rouge, Louisiana 70803/USA

Contents

- I. Introduction 62
- II. Vinyl Mercaptan Precursors 63
  - a) Vinylthioesters 63
  - b) Substituted Styrene Monomers 65
  - c) Vinyl Mercaptals 66
  - d) Vinylthiocarbonates and Carbamates 68
  - e) Nucleophilic Sulfur Reagents 72
- III. Ring Opening Polymerization of Cyclic Monomers 74
  - a) Cysteine Derivatives 75
  - b) Substituted Oxetanes 76
- IV. Mercaptan-Containing Condensation Polymers 77
- V. Chemical Introduction of Mercaptan Groups into Preformed Polymers 78
  - a) Nylon Derivatives 78
  - b) Poly(vinyl alcohol) Derivatives 78
  - c) Polystyrene Derivatives 80
  - d) Chloromethylated Polystyrene Derivatives 82
  - e) Reactions with Bender's Salts 84
- VI. Introduction of Mercaptan Groups into Proteins 85
- VII. References 88

I. Introduction

Mercaptan-containing polymers have been utilized as selective sequestering agents which form stable mercaptides with heavy metals such as Hg, Pb, Ag, Cu, Cd, Fe, Zn, Ni, Pd, and Au. This property has been used for the detection of thiol groups in biological systems as well as for the preparation of exchange resins for the removal of heavy metal ions. Crosslinked poly(vinylbenzyl mercaptan) (*1*) and crosslinked poly(*p*-mercaptostyrene) (*2*) both proved effective as sequestering agents. For example, mercuric ions were removed quantitatively from a nitric acid solution and then selectively recovered by eluting the polymer-captide column with ammonical 2,3-dimercapto-1-propanol (BAL). Several other complexing agents failed to remove the mercury. The selectivity of polymercaptan resins offers many potential applications in analytical and biochemical fields where selective concentration or removal of heavy metals is desired.

The reducing properties of thiols, suggest that mercaptan

$$2RSH \rightleftarrows R-S-S-R + 2H^{\oplus} + 2e^{\ominus},$$

containing polymers would be useful redox resins (*3*). The redox reaction can be reversed with strong reducing agents or monomeric thiols to regenerate the polymercaptan resin. The application of poly(*p*-mercaptostyrene) as a redox resin was explored by Cassidy *et al.* (*4*) but efforts to determine the redox potential were thwarted by precipitation of the polymer during the titration.

Synthetic mercaptan-containing polymers are also interesting in relation to the important role of the thiol function in enzyme systems (*5*) and the catalytic properties of copolymers containing mercaptan and imidazole substituents have been investigated (*6*). Water soluble copolymers have been studied as antiradiation agents (*7*). A short term prophylaxis was observed which was attributed to the ability of mercaptans to inhibit free radical reactions by the formation of stable thiol radicals. A recent ESR study of copolymers of vinyl pyrrolidone and vinyl mercaptan confirms this hypothesis; exposure of these copolymers to ionizing radiation initially produced pyrrolidinyl radicals which rapidly rearranged to exceptionally stable thiyl radicals (*8*).

A variety of synthetic methods exist for the preparation of mercaptan-containing polymers. In contrast to vinyl alcohol, vinyl mercaptan can be isolated in very low yield from the addition of hydrogen sulfide to acetylene in the presence of free radical initiators such as peroxides, azo compounds, and UV light (9). However, it is useless as a monomer because the sulfhydryl

$$HC\equiv CH + H_2S \xrightarrow{h\nu} \underset{\underset{SH}{|}}{CH_2{=}CH}$$

group is too labile, i.e. it is a very efficient chain transfer agent and undergoes trimerization *via* a tautomeric thioacetaldehyde intermediate (*10*). Thus one is forced to resort to indirect synthetic techniques of the following general types:

a) the polymerization of monomers containing a potential mercaptan function which is protected by a subsequently removable "blocking" group,

b) chemical modification of preformed polymers.

Monomer synthesis, followed by polymerization, has the advantage that the composition of the product is known with increased certainty, and that the product is readily characterized. However, the molecular weights of the polymers are usually too low to exhibit good mechanical properties. Chemical modification of preformed high molecular weight polymers should be considered if mechanical properties, total mercaptan content without regard to structural orientation, and minimal cost are of primary importance. The limited solubility and reactive site accessibly of polymeric precursors coupled with extensive side reactions and crosslinking generally cause difficulties in chemical modification of high polymers (*11–14*). Thus, the selection of an appropriate synthesis of mercaptan containing polymer, i.e. from monomer, or from preformed polymer, clearly depends upon the ultimate application of the product.

II. Vinyl Mercaptan Precursors

a) Vinylthioesters

The application of several different blocking groups to polymercaptan synthesis has been thoroughly evaluated. In general, this technique enables one to prepare homopolymers of vinyl mercaptan under conditions where the structure of the polymer can be assigned accurately. For example, monomers containing thioester functional groups have been studied extensively. S-vinyl thioacetate (II, R=CH$_3$) was prepared in 43% yield by cracking the diacetate of mercaptoethanol (I) at 500° C (15). Recent attempts to duplicate these results revealed that the yields are extremely variable and that the by-products are powerful vesicants.

$$CH_2=CH-O-\overset{\overset{O}{\|}}{C}-CH_3 + R-\overset{\overset{O}{\|}}{C}-SH \xrightarrow[\text{or } H_2SO_4]{O_2} R-\overset{\overset{O}{\|}}{C}-S-CH_2CH_2O-\overset{\overset{O}{\|}}{C}-CH_3$$
$$(R=CH_3, C_6H_5) \qquad\qquad (I)$$

$$CH_2=CH \qquad\qquad\qquad \Big| \;\; 520\text{–}550°\text{ C}$$
$$\;\;\;\;\;\;\;\;\;\;\;\; | \qquad\qquad\qquad\quad -CH_3COOH$$
$$\;\;\;\;\;\;\;\;\;\;\;\; S-\underset{\underset{O}{\|}}{C}-R \longleftarrow$$

(II)

The polymerization and copolymerization of vinyl thioacetate in the presence of radical catalysts does not lead to high molecular weight polymers because II exhibits a high chain transfer constant (5.6×10^{-2}) (18). Subsequent hydrolysis of the thioacetate functional group requires alkaline conditions and is accompanied by oxidative crosslinking (17). Polymerization of vinyl thioacetate in liquid ammonia yields poly(vinyl thiol) with intra- or intermolecular disulfide bridges, the acetate part of the molecule is concomitantly split off to form acetamide (22). Copolymerization of II with vinylene carbonate (16), methylmethacrylate (15, 20), N-vinylsuccinimide (18), N-vinylphthalimide (18), N-vinylcarbazole (18), styrene (15, 19), vinylacetate (15, 19), and methyl acrylate (19) as well as the hydrolysis of these copolymers, has been reported. Copolymers of vinyl thioacetate with esters of 1,4-butenedioic acid can be used as coating compounds; their hydrolyzed derivatives can be used in cationic-exchange resins and water softeners (21).

In a similar fashion, vinyl thiobenzoate was synthesized in low yield (~10%) by pyrolysis of β-acetoxyethyl thiobenzoate (I, R=C$_6$H$_5$) which was derived from thiobenzoic acid and vinyl acetate in the presence of sulfuric acid (23). The polymerization and copolymerization of II (R=C$_6$H$_5$) were carried out in bulk using azobisisobutyronitrile (AIBN) as initiator. Basic hydrolysis of the polymer yielded poly(vinyl mercaptan), but no sufficient improvement in the degree of polymerization was reported.

b) Substituted Styrene Monomers

Poly(p-mercaptostyrene) was prepared by the polymerization of p-vinylphenyl thioacetate (24, 25) (III) which was synthesized from p-aminoacetophenone (IV) (Scheme I). Conversion of IV *via* a diazonium salt to xanthate (V), followed by reduction with $NaBH_4$ introduced a thiophenol group (VI). Pyrolysis of the corresponding diacetate produced III in ~10% overall yield. This styrene derivative polymerized readily and saponification of the resultant homopolymer yielded an alkali soluble polymer. Copolymerization with methyl methacrylate gave, upon saponification, another synthetic mercaptan-containing copolymer.

An interesting approach to polymercaptans utilizes the mercaptide ion as a blocking group. Vinylbenzyl thioacetate (26) (IX) was prepared in 68.5% yield by treating vinylbenzyl chloride (VIIIa) (70:30 p- to o-isomer) with potassium thioacetate. Basic hydrolysis of IX yielded

[Scheme showing synthesis of vinylbenzyl mercaptan:
- VIIIa,b (CH₂=CH-C₆H₄-CH₂Cl) reacts with CH₃C(O)SH/KOH to give IX (CH₂=CH-C₆H₄-CH₂-S-C(O)-CH₃)
- IX treated with (i) KOH/CH₃OH (ii) HCl/H₂O gives XIa,b (CH₂=CH-C₆H₄-CH₂SH)
- VIIIa,b reacts with H₂NC(S)NH₂ to give X (CH₂=CH-C₆H₄-CH₂-S-C(=NH·HCl)-NH₂)
- X treated with NH₃/H₂O gives XIa,b]

42.6% of the corresponding vinylbenzyl mercaptan (XIa). Alternatively, S-(4-vinylbenzyl)-isothiuronium chloride (27) (X) was prepared by treating pure 4-vinylbenzyl chloride (VIIIb) with thiourea. Subsequent hydrolysis of X yielded 78% 4-vinylbenzyl mercaptan (XIb).

Heating XIa with AIBN for four days at 100° yielded a poly(vinylbenzyl mercaptan) with 16.6 wt.-% of sulfur. Emulsion polymerization of XIb initiated by potassium persulfate in alkaline media yielded the corresponding insoluble mercaptan-containing polymer. Alkali metal salts of the polymer from XIa are water soluble and may be oxidized with $Ca(OCl)_2$ to give insoluble products containing disulfide linkages between the polymer chains. Detailed characterization of these polymers was not reported, so the efficiency of the mercaptide ion in preventing chain transfer reactions could not be evaluated. However, it is difficult to conceive of a simpler blocking group.

c) Vinyl Mercaptals

An alternate approach to the synthesis of polymercaptan precursors was considered by H. Ringsdorf et al. (28–31). They prepared vinyl mercaptals by either dehydration of the analogues S-β-hydroxyethyl compounds or dehydrohalogenation of S-(β-chloroethyl)monothioacetals (Scheme II). The monomers were polymerized radically with AIBN and cationically with BF$_3$-etherate but only low molecular weight polymers were obtained regardless of the reaction conditions. Further, the polymeric mercaptal could not be hydrolyzed to polymercaptan.

Preparation and cyclopolymerization of S,S'-divinylmercaptals (XII) have also been reported by two different groups. Ringsdorf and Overberger (28, 31) found that S,S'-divinyldithioformal (XII, X=S; R = —CH=CH$_2$) can be prepared by dehydration of S,S'-(β-hydroxyethyl)dithioformal. Free radical polymerization of (XII) in dilute solution to low conversions favored the formation of linear polydivinylmercaptal containing dithiane repeat units (XIII). The extremely low molecular weight soluble portion contained approximately 26% residual unsaturation and crosslinked rapidly upon exposure to air (32). Matsoyan and Soakyan (32) prepared S,S'-divinyldithioformal by treatment of ethylene sulfide with hydrogen chloride in the presence of paraformalde-

Scheme II

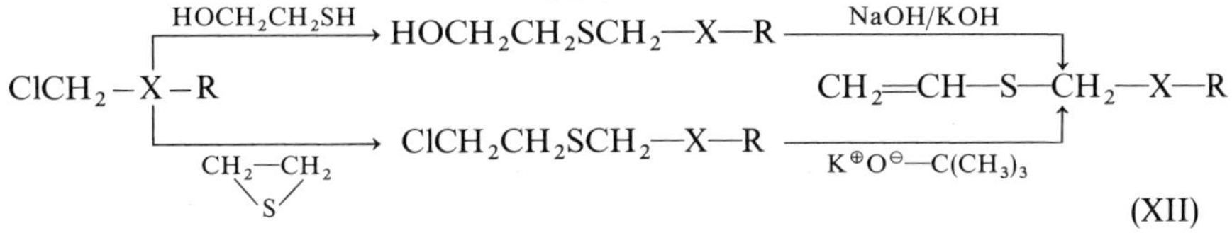

$$ClCH_2-X-R \xrightarrow{HOCH_2CH_2SH} HOCH_2CH_2SCH_2-X-R \xrightarrow{NaOH/KOH} CH_2{=}CH-S-CH_2-X-R$$

$$ClCH_2-X-R \xrightarrow{\overset{CH_2-CH_2}{\underset{S}{\diagdown\diagup}}} ClCH_2CH_2SCH_2-X-R \xrightarrow{K^\oplus O^\ominus-C(CH_3)_3}$$

(XII)

where

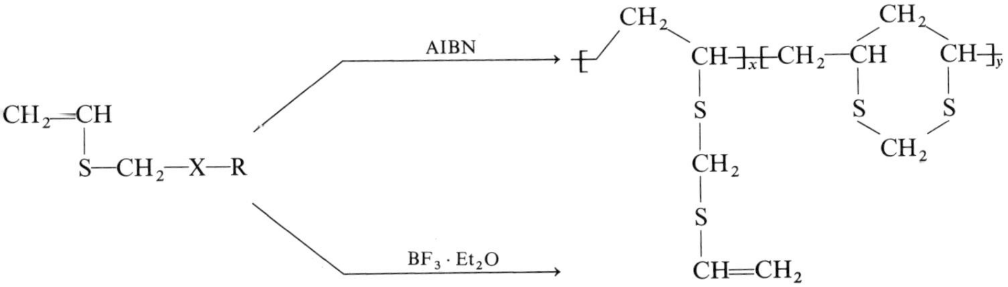

$$\begin{bmatrix} X = S, O. \\ R = CH_3-,\ CH_3CH_2-,\ \bigcirc,\ -CH\overset{CH_3}{\underset{CH_3}{\diagup\diagdown}},\ -CH_2CH_2Cl,\ -CH{=}CH_2. \end{bmatrix}$$

$$CH_2{=}CH\underset{S-CH_2-X-R}{\Big|} \xrightarrow[BF_3\cdot Et_2O]{AIBN} $$

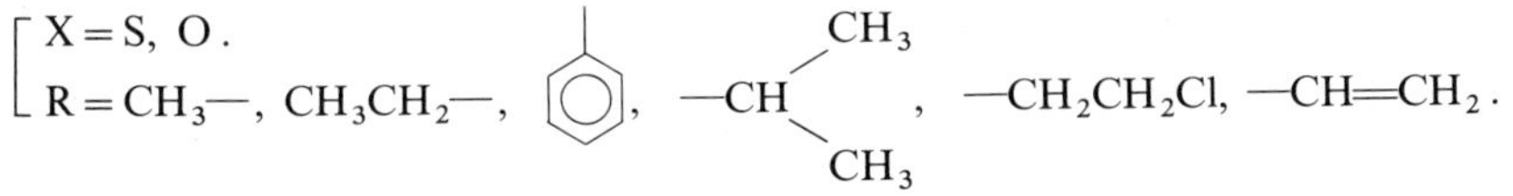

(XIII)

hyde to produce S,S'-(β-chloroethyl)dithioformal which was subsequently dehydrogenated. Similarly S,S'-divinyldithioacetal and S,S'-divinyldithiopropional were synthesized from the corresponding aldehydes. The substituted mercaptals cyclopolymerized in bulk but soluble poly(S,S'-divinyldithioformal) could only be obtained by allowing the monomer to stand in the sun for eight months.

The extensive 1,3-dithiane studies by Corey et al. (33) showed that one should be able to alkylate polymers containing dithiane rings (XIII) by preparing a lithio salt of the polymer with butyl lithium followed by treatment with an alkyl halide, but all attempts from this laboratory to release the elaborated aldehydes generated by this process failed.

d) Vinylthiocarbonates and Carbamates

The preparation and cyclopolymerization of S,S'-divinyldithiocarbonate has been reported by Ringsdorf and Overberger (34). Their results indicated that this could be a potential monomer for the preparation of stereoregular polymercaptans. Reinvestigation of the monomer synthesis conducted later by Overberger and Daly showed that dehydrohalogenation of S,S'-(β-chloroethyl)-dithiocarbonate (XIV) with potassium t-butoxide yielded a mixture of S,S'-divinylthiocarbonate (XV), S-vinyl-O-t-butylthiocarbonate (XVI), S-(β-chloroethyl)-O-t-butylthiocarbonate (XVII), S-(β-vinylmercaptoethyl)-S'-vinyl-dithiocarbonate (XVIII), and S-vinyl-S'-(β-chloroethyl)dithiocarbonate (XIX) (35). The complexity of the reaction mixture is an example of the facile rearrangements that thiocarbonates undergo in basic media. The cyclopolymers initially reported (34) were derived from (XVIII) but free radical copolymers of XV also yielded cyclopolymers of low molecular weight ($[\eta] = 0.05$) (36).

$$\begin{array}{c}\text{ClCH}_2\text{CH}_2\text{S}\\ \diagdown\\ \text{C}=\text{O}\\ \diagup\\ \text{ClCH}_2\text{CH}_2\text{S}\end{array} \xrightarrow{t\text{-BuOK}} \begin{array}{c}\text{CH}_2=\text{CHS}\\ \diagdown\\ \text{C}=\text{O}\\ \diagup\\ \text{CH}_2=\text{CHS}\end{array} + \text{CH}_2=\text{CH}-\text{S}-\underset{\underset{\text{O}}{\|}}{\text{C}}-\text{O}-\text{C}(\text{CH}_3)_3$$

(XIV)   (XV)   (XVI)

$$+ \text{ClCH}_2\text{CH}_2\text{S}-\underset{\underset{\text{O}}{\|}}{\text{C}}-\text{O}-\text{C}(\text{CH}_3)_3 + \text{CH}_2=\text{CH}-\text{S}-\underset{\underset{\text{O}}{\|}}{\text{C}}-\text{S}-\text{CH}_2\text{CH}_2\text{Cl} +$$

(XVII)   (XIX)

$$+ \text{CH}_2=\text{CH}-\text{S}-\text{CH}_2\text{CH}_2\text{S}-\underset{\underset{\text{O}}{\|}}{\text{C}}-\text{S}-\text{CH}=\text{CH}_2$$

(XVIII)

Since cystamine is one of the most effective radiation prophylactics known, the preparation and polymerization of S-vinyl-N-vinylthiocarbamates (37) (XXI) was utilized in an attempt to prepare alternating poly(vinylamine-vinylmercaptan)copolymers (Scheme III). The divinyl monomers were prepared by dehydrochlorination of the respective S-2-chloroethyl-N-2-chloroethylthiocarbamates (XX). Polymerization of these monomers under a variety of conditions yielded "terpolymers" (XXII) composed of the tetrahydro-1,3-thiazin-2-one moiety as well as poly-S-vinyl and poly-N-vinyl units. Attempts to isolate polymeric α-amino γ-thiols (XXIII) formed by hydrolysis of the resultant "terpolymers" (XXII) proved unsuccessful.

In conjunction with this investigation of sulfur containing polymers as potential radiation prophylactics, Overberger et al. (38) synthesized S-(β-chloroethyl)thiochloroformate (XXIV) and utilized it to prepare several vinyl thiocarbonates and thiocarbamates. These monomers were polymerized by free radical initiation and several yielded reasonably high molecular weight polymers, but the polymers proved difficult to hydrolyze to the desired polymercaptans. However, utilization of the labile carbo-t-butoxy blocking group in the monomer synthesis by

Scheme III

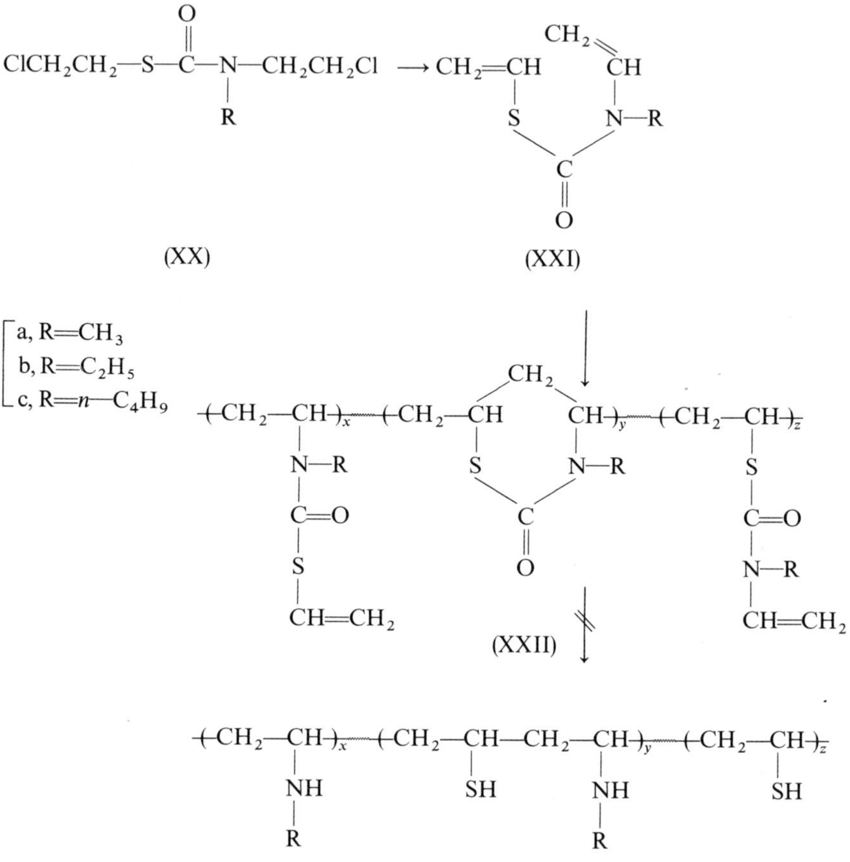

ClCH$_2$CH$_2$—S—C(=O)—N(R)—CH$_2$CH$_2$Cl → structure XXI (cyclic: CH$_2$=CH—S—C(=O)—N(R)—CH—CH$_2$)

(XX) (XXI)

a, R=CH$_3$
b, R=C$_2$H$_5$
c, R=n—C$_4$H$_9$

+(CH$_2$—CH)$_x$—(CH$_2$—CH)—(CH)$_y$—(CH$_2$—CH)$_z$
with substituents: N—R / C=O / S / CH=CH$_2$; ring (S—C(=O)—N—R bridging CH$_2$); S / C=O / N—R / CH=CH$_2$

(XXII)

↓

+(CH$_2$—CH)$_x$—(CH$_2$—CH—CH$_2$—CH)$_y$—(CH$_2$—CH)$_z$
| | | |
NH   SH   NH   SH
| |
R   R

(XXIII)

preparing S-vinyl-O-*t*-butylthiocarbonate alleviated the hydrolysis problems and produced an excellent monomeric precursor for polymercaptans (*39*). The blocking group could be removed by treatment of the polymer with anhydrous HBr in 6:1 chloroform-tetrachloroethane or by thermolysis at 150° in N-methylpyrrolidone; conditions which minimized the oxidative side reactions that accompany the liberation of free mercaptans in an alkaline media. The autoxidation of these polymers and related model compounds was studied in an effort to ascertain the influence of the polymer structure on reactivity (*40*).

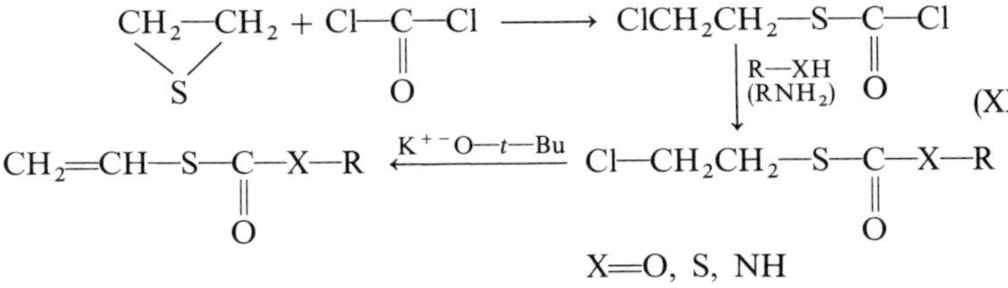

$$\underset{S}{CH_2-CH_2} + Cl-\underset{O}{\overset{\|}{C}}-Cl \longrightarrow ClCH_2CH_2-S-\underset{O}{\overset{\|}{C}}-Cl$$

$$\downarrow \begin{array}{c} R-XH \\ (RNH_2) \end{array}$$ (XXIV)

$$CH_2=CH-S-\underset{O}{\overset{\|}{C}}-X-R \xleftarrow{K^+ \ ^-O-t-Bu} Cl-CH_2CH_2-S-\underset{O}{\overset{\|}{C}}-X-R$$

X=O, S, NH

S-vinyl-O-*t*-butylthiocarbonate (XVI) polymerizes readily in the presence of free radical initiators to yield polymers with a maximum molecular weight of 55,000 (DP ≅ 335). Although this is an order of magnitude higher than the molecular weight attainable with vinyl thioacetate under similar conditions, XVI exhibits a chain transfer constant of 3.9×10^{-3} which limits the molecular weight of both homopolymers and copolymers prepared under free radical conditions. The copolymerization characteristics of S-vinyl-O-*t*-butylthiocarbonate were typical of vinylthiocarbonyl monomers; the Q and e copolymerization parameters are consistent with a resonance stabilized monomer and radical with a negatively polarized double bond (*41*). Water soluble copolymers of XVI can be prepared by incorporating 60% vinyl pyrrolidone but the hydrophobicity of free mercaptan groups renders these copolymers insoluble in water upon removal of the blocking group; a copolymer of vinyl mercaptan containing 95% vinylpyrrolidone remains water soluble.

One of the problems encountered in working with S-vinyl-O-*t*-butylthiocarbonate was a tendency for this monomer to yield crosslinked polymer. The crosslinking was apparently due to the presence of 3–5% S,S'-divinyldithiocarbonate (XV) formed by the reaction sequence shown in Scheme IV.

Scheme IV

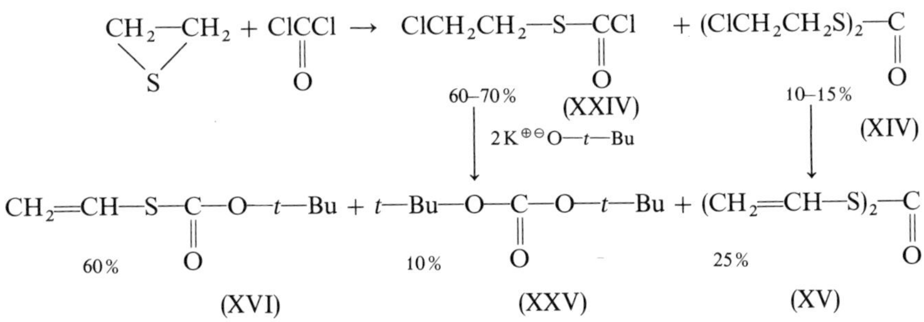

$$\underset{S}{CH_2-CH_2} + Cl_3CCl \rightarrow \underset{\underset{60-70\%}{(XXIV)}}{ClCH_2CH_2-\underset{O}{\overset{\|}{S-CCl}}} + \underset{10-15\%}{(ClCH_2CH_2S)_2-\underset{O}{\overset{\|}{C}}}$$

$$\downarrow 2K^{\oplus \ \ominus}O-t-Bu \qquad \qquad \downarrow \text{(XIV)}$$

$$\underset{60\%}{CH_2=CH-S-\underset{O}{\overset{\|}{C}}-O-t-Bu} + \underset{10\%}{t-Bu-O-\underset{O}{\overset{\|}{C}}-O-t-Bu} + \underset{25\%}{(CH_2=CH-S)_2-\underset{O}{\overset{\|}{C}}}$$

(XVI) (XXV) (XV)

It was very difficult to prevent multiple addition of ethylenesulfide to phosgene, and the tendency of (XXIV) and (XIV) to decompose during distillation made separation of the two products difficult.

e) Nucleophilic Sulfur Reagents

Okawara *et al.* (*42*), reported that the addition of sodium N,N-dialkyldithiocarbamate to 1,2-dichloroethane (Scheme V) in DMF yielded a diadduct (XXVI) which could be cracked to produce S-vinyl-N,N-dialkyldithiocarbamate (XXVII). Monomer (XXVII) could be polymerized under free radical conditions, but it exhibited a high chain transfer constant so the molecular weight of the resultant polymer (XXVIII) was rather low. Furthermore, the thermal stability of the polymer was poor, which is not surprising since the polymer structure was very similar to diadduct (XXVI), which could be cracked at relatively low temperatures. Polymercaptan (XXIX) could be generated from (XXVIII) by treatment with dimethylamine. Polymers containing the dithiocarbamate moiety are reported to be effective photosensitive resins (*11, 43*).

Application of similar nucleophilic displacement reactions to the preparation of S-vinyl-O-alkylthiocarbonates simply required the synthesis of appropriate Bender's salts. Bender's salts are stable crystalline

Scheme V

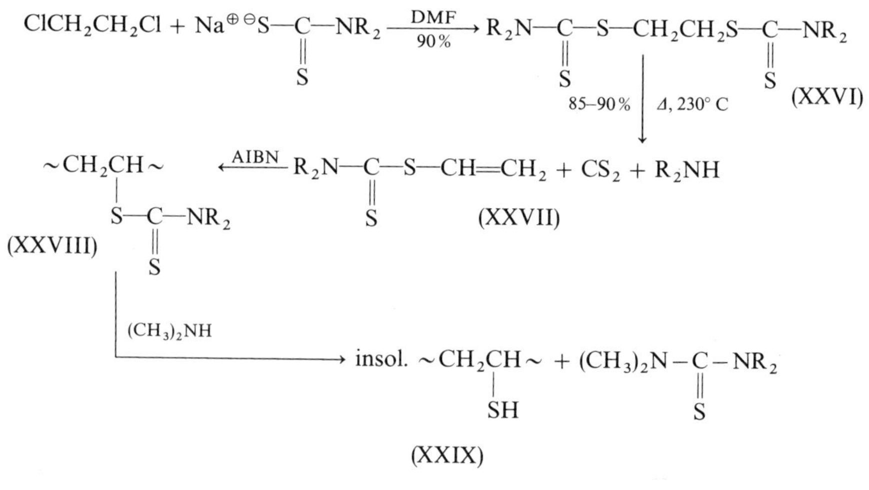

$$ClCH_2CH_2Cl + Na^{\oplus \ominus}S-\underset{\underset{S}{\|}}{C}-NR_2 \xrightarrow[90\%]{DMF} R_2N-\underset{\underset{S}{\|}}{C}-S-CH_2CH_2S-\underset{\underset{S}{\|}}{C}-NR_2$$

85–90% | Δ, 230°C (XXVI)

$$\sim CH_2CH\sim \quad \xleftarrow{AIBN} \quad R_2N-\underset{\underset{S}{\|}}{C}-S-CH=CH_2 + CS_2 + R_2NH$$

(XXVIII) S—C—NR₂ ‖ S   (XXVII)

(CH₃)₂NH → insol. ~CH₂CH~ + (CH₃)₂N—C—NR₂
                        | ‖
                        SH   S

(XXIX)

reagents which can be prepared by treating potassium alcoholates with carbonyl sulfide in an anhydrous solvent mixture of alcohol and dimethylformamide (DMF) (44). In contrast to potassium xanthates, the salts are insoluble in most organic solvents but a mixture of 1–2% water in DMF is a good solvent system for homogenous displacement reactions. In the dry form, the salts are quite stable and samples have been stored over a year in an argon atmosphere without noticeable loss of reactivity. As Scheme VI illustrates, selective displacement of bromide in the presence of chloride can be effected in THF: *t*-butanol (4:1) at −10°. Subsequent dehydrohalogenation of the S-(β-chloroethyl)-O-*t*-butylthiocarbonate (XVIIa) produced monomer (XVIa) in 58% overall yield. The monomer was still contaminated by 2–3% di-*t*-butylcarbonate but this appeared to act as an inert diluent in the polymerizations. Using potassium O-*t*-amylthiocarbonate (XXXb) as the nucleophile, S-vinyl-O-*t*-amylthiocarbonate (XVIb) could be prepared *via* Scheme VI in 51% yield.

Scheme VI

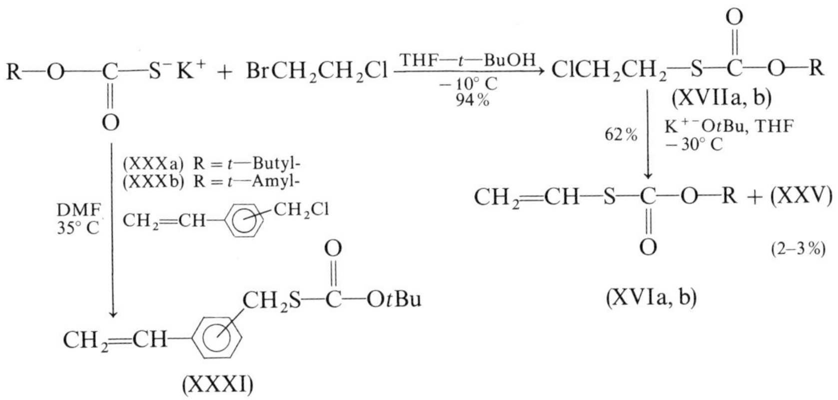

$$R-O-\underset{\underset{O}{\|}}{C}-S^-K^+ + BrCH_2CH_2Cl \xrightarrow[-10°C]{THF-t-BuOH}_{94\%} ClCH_2CH_2-\underset{\underset{O}{\|}}{S-C}-O-R$$

(XVIIa, b)

(XXXa) R = *t*—Butyl-
(XXXb) R = *t*—Amyl-

$$\downarrow \text{DMF, 35°C} \quad CH_2=CH-\text{⟨Ph⟩}-CH_2Cl$$

$$62\% \downarrow K^{+-}OtBu, THF, -30°C$$

$$CH_2=CH-\underset{\underset{O}{\|}}{S-C}-O-R + (XXV)$$

(XVIa, b) (2–3%)

$$CH_2=CH-\text{⟨Ph⟩}-CH_2S-\underset{\underset{O}{\|}}{C}-OtBu$$

(XXXI)

An alternative approach to a monomer synthesis utilized a commercially available mixture of 60% *meta* and 40% *para*-vinylbenzyl chloride as the substrate. Treatment of the mixture with XXXa in DMF produced S-vinylbenzyl-O-*t*-butylthiocarbonate (XXXI) in quantitative yield. Polymerization of XXXI under free radical conditions yielded a mixture of benzene soluble and insoluble crosslinked polymer. Thermolysis of the soluble fraction in N-methylpyrrolidone generated a soluble poly(vinylbenzyl mercaptan).

Recently the synthesis of polymers and copolymers of isothiuronium salts was reported (45). The authors prepared 2-[(methacryloxy)-alkyl]isothiuronium salts (XXXII) by the following procedure:

$$\begin{array}{c}CH_2{=}C{-}CH_3\\|\\C{=}O\\|\\O{-}(CH_2)_n{-}OH\\n=2,3,4\end{array} + \begin{array}{c}SO_2Cl\\|\\\bigcirc\\|\\CH_3\end{array} \longrightarrow \begin{array}{c}CH_2{=}C{-}CH_3\\|\\C{=}O\\|\\O{-}(CH_2)_n{-}OTs\\\downarrow\ NH_2{-}C({=}S){-}NH_2\end{array}$$

$$\xrightarrow{\text{AIBN, }t\text{-BuOH}}\begin{array}{c}CH_2{=}C{-}CH_3\\|\\C{=}O\\|\\O{-}(CH_2)_n{-}S{-}C({=}NH_2^+){-}NH_2\quad OTs^-\end{array}\quad(XXXII)$$

20–25% Polymer
↓ NaOH

$$\sim CH_2{-}\overset{\text{\large ?}}{C}{-}CH_3 \qquad + (CH_2{-}CH_2)\\\phantom{\sim CH_2{-}}|\phantom{{-}CH_3}\qquad\qquad\ \ \diagdown S\diagup\\\phantom{\sim CH_2{-}}C{=}O\\\phantom{\sim CH_2{-}}|\\\phantom{\sim CH_2{-}}O{-}(CH_2)_n{-}S^-$$

XXXII could be polymerized or copolymerized with other monomers, e.g. acrylic acid and acrylamide. Polymerization of these monomers proceeded very slowly even in polar solvents and high degrees of conversion were difficult to obtain. Alkaline hydrolysis of the isothiuronium salt was accompanied by ethylene sulfide evolution in some cases; this side reaction reduced the yield of mercaptide groups significantly. The polymers may be useful as antifogging agents and silver photographic-image stabilizers. A similar approach was used to make vinyl isothiouronium salts and their copolymers (46).

III. Ring Opening Polymerization of Cyclic Monomers

Since the chain transfer properties of vinyl mercaptan precursors limited the molecular weights attained in addition polymerization, several ring-opening polymerizations of monomers with thio-substituents were evaluated.

a) Cysteine Derivatives

Poly-L-cysteine (XXXIV) was prepared by polymerization of S-carbobenzoxy-N-carboxy-L-cysteine anhydride (XXXIII), followed by reduction with sodium in liquid ammonia (47). The reactivity of the thiol groups in this polymer toward iodoacetic acid indicated that quantitative liberation of mercapto groups had been effected but selective reactivity was observed with methyl mercuric nitrate in the presence of sodium nitroprusside, an internal indicator. These reactivity differences result from steric exclusion of the nitroprusside dye from about one third of the available mercaptan moities. Thus the concept of "sluggish-SH groups" in a protein has been demonstrated with a synthetic homopolymer.

$$\begin{array}{c} \text{NH—COOCH}_2\text{C}_6\text{H}_5 \\ | \\ \text{CH—COOH} \\ | \\ \text{CH}_2\text{—S—COOCH}_2\text{C}_6\text{H}_5 \end{array} \xrightarrow{\text{PCl}_5} \begin{array}{c} \text{NH—CO} \\ | \quad\quad\;\;\diagdown\text{O} \\ \text{CH—CO}\diagup \\ | \\ \text{CH}_2\text{—S—COOCH}_2\text{—C}_6\text{H}_5 \\ \text{(XXXIII)} \end{array} \xrightarrow{-\text{CO}_2} \begin{array}{c} \text{H}\text{—}\!\!\left[\text{NH—CH—CO}\right]_{\!x}\!\!\text{—OH} \\ | \\ \text{CH}_2 \\ | \\ \text{S—COOCH}_2\text{C}_6\text{H}_5 \end{array}$$

$$\begin{array}{c} \text{H}\text{—}\!\!\left[\text{NH—CH—CO}\right]_{\!x}\!\!\text{—OH} \\ | \\ \text{CH}_2 \\ | \\ \text{SH} \\ \text{(XXXIV)} \end{array} \xleftarrow{\text{Na, liq. NH}_3}$$

Poly-(N-methacrylyl cysteine) has been prepared by using S-thiophenyl or S-benzyl blocking groups to protect the thiol substituent during methacrylation and polymerization (48). Treatment of the resultant polymer with sodium in liquid ammonia produced free cysteine residues in 51% yield. Alternatively, N,N'-*bis*-methacrylyl cystine was prepared, polymerized to a crosslinked gel and 79% of the theoretical cysteine residue were liberated on reduction of the disulfide linkage. These polymers should be of interest as a potential radiation prophylatics.

$C_6H_5-S-S-CH_2-CH(NH_2)-COOH \longrightarrow C_6H_5-S-S-CH_2-CH(NH-C(=O)-C(CH_3)=CH_2)-COOH$

[Scheme showing polymerization: from the methacrylamide monomer to polymer $\{CH_2-CH\}_x$ with CH$_3$ side group, C=O, NH-CH(CH$_2$S-S-C$_6$H$_5$)-COOH; then to polymer $\{CH_2-C\}_x$ with CH$_3$, C=O, NH-CH(CH$_2$SH)-COOH]

b) Substituted Oxetanes

In an effort to improve the mechanical properties of polymercaptans, Goethals turned to cationic polymerization of 3,3-*bis*-(methylenethioester)oxetanes (*49*). Using dithiocarbonates or *bis*-(thioacetate) derivatives (XXXV), polymers with molecular weights as high as 21,100 could be obtained. Higher molecular weight materials were prepared by copolymerizing the sulfur containing monomers with 3,3-*bis*-(chloromethyl)oxetane but extensive crosslinking occurred during the deblocking step due to the nucleophilic attack of liberated mercaptide ions on residual chloromethyl substituents.

[Reaction scheme: 3,3-bis(chloromethyl)oxetane + Na$^+$ $^-$S-C(=O)-CH$_3$ → 3,3-bis(CH$_2$SCCH$_3$(=O))oxetane (XXXV); then BF$_3\cdot$Et$_2$O / CH$_2$Cl$_2$ polymerization to polymer with CH$_2$-S-C(=O)-CH$_3$ side groups, $\bar{M}n = 21,100$; then 1. NaOH, 2. HCl to give polymer with SH SH groups, only sol. in alkaline media]

IV. Mercaptan-Containing Condensation Polymers

Synthetic methods for the preparation of blocked sulfhydryl monomers and their subsequent condensation polymerization were illustrated by the work of Overberger and Aschkenasy (50). Employing the carbobenzoxy blocking group, a mercaptan-containing polyamide (XXXVII) of α,α-dithioladipyl chloride (XXXVI) and hexamethylenediamine was prepared. Similarly, a polyurethane (XXXIX) synthesized from 2-benzylthiomethyl-1,4-butanediol (XXXVIII) and hexamethylene diisocyanate yielded a polymercaptan on debenzylation.

$$\underset{(XXXVI)}{\underset{\underset{COOCH_2C_6H_5}{|}}{\underset{S}{|}}{ClC}(=O)-CH-(CH_2)_2-\underset{\underset{COOCH_2C_6H_5}{|}}{\underset{S}{|}}{CH}-C(=O)-Cl} + H_2N(CH_2)_6NH_2 \xrightarrow[\text{with HBr/CF}_3\text{COOH}]{\text{after treatment}} \underset{(XXXVII)}{[-C(=O)-\underset{\underset{SH}{|}}{CH}-(CH_2)_2-\underset{\underset{SH}{|}}{CH}-C(=O)-NH-(CH_2)_6-NH-]_x}$$

$$\underset{(XXXVIII)}{\underset{\underset{S-CH_2-C_6H_5}{|}}{\underset{CH_2}{|}}{HO-CH_2-CH-(CH_2)_2-OH}} + OCN-(CH_2)_6-NCO \xrightarrow[\text{Na/liq. NH}_3\text{-propylamine}]{\text{after treatment}} \underset{(XXXIX)}{[-O-CH_2-\underset{\underset{SH}{|}}{\underset{CH_2}{|}}{CH}-(CH_2)_2-O-C(=O)-NH-(CH_2)_6-NH-C(=O)-]_x}$$

Polycondensation of thiophenyl and derivatives of thioglycolic acid anilide with formaldehyde yielded resins with the free thiophenol functions. The oxidation potentials of the polymers were obtained indirectly by Hamamura et al. (51) by observing the ability of the polymers to decolorize redox dyes of known potentials.

V. Chemical Introduction of Mercaptan Groups into Preformed Polymers

a) Nylon Derivatives

Cairns and co-workers (52) introduced a functional group into poly(hexamethylene adipamide) (Nylon 66) by hydroxymethylating the polyamide backbone with formaldehyde and ammonia. Subsequent treatment with thiourea and hydrogen chloride gave an isothiuronium salt from which an alkali soluble polythiol could be obtained.

$$\begin{array}{c}\text{-[NH-(CH}_2)_6\text{-NH-CO-(CH}_2)_4\text{-CO-]}_x \xrightarrow{\text{HCHO/NH}_3} \text{-[N(CH}_2\text{OH)-(CH}_2)_6\text{-N(CH}_2\text{OH)-CO-(CH}_2)_4\text{-CO-]}_x \\ \xrightarrow{\text{NH}_2\text{CSNH}_2 / \text{HCl}} \xrightarrow{\text{H}_2\text{O}} \text{-[N(CH}_2\text{SH)-(CH}_2)_6\text{-N(CH}_2\text{SH)-CO-(CH}_2)_4\text{-CO-]}_x\end{array}$$

Okawara et al. (53) reinvestigated this procedure and found that thiourea treatment followed by saponification yielded a polymercaptan containing only 0.61% SH groups. The low conversion observed in this process was probably due to a decrease on the functional groups on the polymer since hydrolysis of the hydroxymethyl substituent also occurs readily under acidic conditions.

b) Poly(vinyl alcohol) Derivatives

Mercaptan-containing polymers derived from poly(vinyl alcohol) (XL) were prepared in a similar manner by treating XL with hydrogen bromide and thiourea to give the isothiuronium salt which could be hydrolyzed to an insoluble polymer of vinyl mercaptan (54, 55).

Poly(vinyl mercaptan) was also made by benzene sulfonation of XL with PhSO$_2$Cl to activate the polyol before thiourea treatment or thioesterification with MeCOSK and hydrolysis of the thioester (56) (Scheme VII). The extent of incorporation of free mercaptan groups was approximately 42%.

Scheme VII

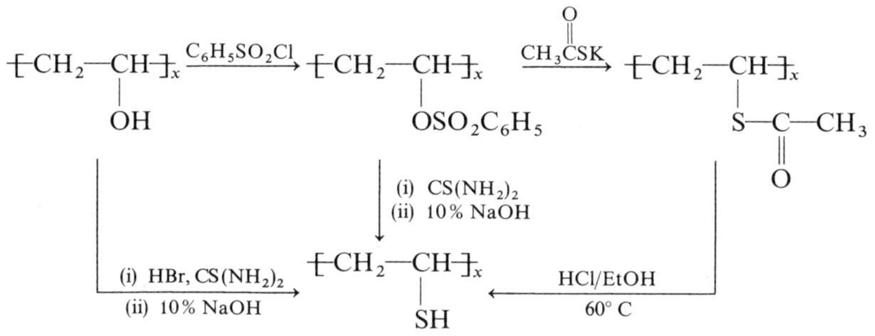

$$\text{+CH}_2\text{—CH+}_x \xrightarrow{\text{C}_6\text{H}_5\text{SO}_2\text{Cl}} \text{+CH}_2\text{—CH+}_x \xrightarrow{\text{CH}_3\overset{\overset{\text{O}}{\|}}{\text{C}}\text{SK}} \text{+CH}_2\text{—CH+}_x$$
$$\quad\quad\quad |\quad\quad\quad\quad\quad\quad\quad\quad |\quad\quad\quad\quad\quad\quad\quad\quad |$$
$$\quad\quad\quad\text{OH}\quad\quad\quad\quad\quad\quad\text{OSO}_2\text{C}_6\text{H}_5 \quad\quad\quad\quad \text{S—C—CH}_3$$
$$\quad\quad\quad\quad\quad\quad\quad\quad\quad\quad\quad\quad\quad\quad\quad\quad\quad\quad\quad\quad\quad\quad\quad\quad \overset{\|}{\text{O}}$$

(i) CS(NH$_2$)$_2$
(ii) 10% NaOH

(i) HBr, CS(NH$_2$)$_2$ +CH$_2$—CH+$_x$ HCl/EtOH
(ii) 10% NaOH | 60° C
                                              SH

Treatment of poly(vinyl alcohol) with chloroacetaldehyde gave a cyclic acetal with a replaceable chloride (XLI). Reaction with KSH or thiourea followed by saponification gave a polythiol (53, 56) (XLII). This approach could lead to hydrophilic polymers with readily accessible mercaptan groups if the initial acetalization is limited to ~10% reaction. Polymers of this type would be useful as enzyme supports.

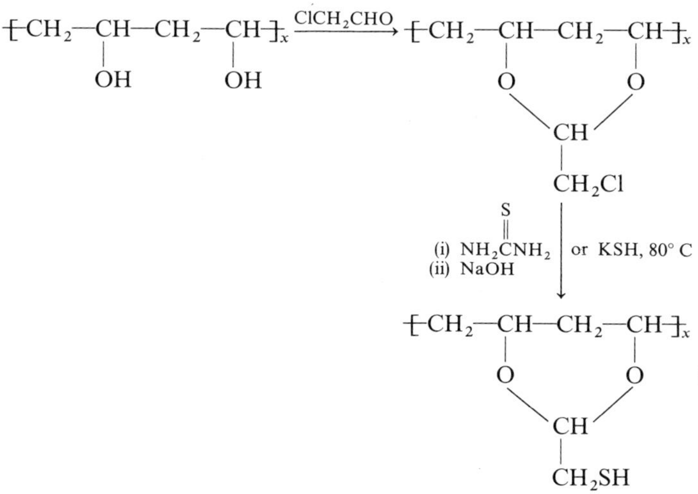

$$\text{+CH}_2\text{—CH—CH}_2\text{—CH+}_x \xrightarrow{\text{ClCH}_2\text{CHO}} \text{+CH}_2\text{—CH—CH}_2\text{—CH+}_x$$
$$\quad\quad\quad\quad\quad |\quad\quad\quad\quad |\quad\quad\quad\quad\quad\quad\quad\quad\quad\quad\quad\quad |\quad\quad\quad\quad\quad |$$
$$\quad\quad\quad\quad\text{OH}\quad\quad\text{OH}\quad\quad\quad\quad\quad\quad\quad\quad\quad\text{O}\quad\quad\quad\quad\text{O}\quad\quad\text{(XLI)}$$
$$\quad\quad\quad\quad\quad\quad\quad\quad\quad\quad\quad\quad\quad\quad\quad\quad\quad\quad\quad\quad\quad\quad\searrow\quad\swarrow$$
$$\quad\quad\quad\quad\quad\quad\quad\quad\quad\quad\quad\quad\quad\quad\quad\quad\quad\quad\quad\quad\quad\quad\quad\text{CH}$$
$$\quad\quad\quad\quad\quad\quad\quad\quad\quad\quad\quad\quad\quad\quad\quad\quad\quad\quad\quad\quad\quad\quad\quad\;\;|$$
$$\quad\quad\quad\quad\quad\quad\quad\quad\quad\quad\quad\quad\quad\quad\quad\quad\quad\quad\quad\quad\quad\quad\text{CH}_2\text{Cl}$$

(i) NH$_2\overset{\overset{\text{S}}{\|}}{\text{C}}NH_2$ or KSH, 80° C
(ii) NaOH

$$\text{+CH}_2\text{—CH—CH}_2\text{—CH+}_x$$
$$\quad\quad\quad |\quad\quad\quad\quad\quad |$$
$$\quad\quad\text{O}\quad\quad\quad\text{O}\quad\quad\text{(XLII)}$$
$$\quad\quad\quad\searrow\;\swarrow$$
$$\quad\quad\quad\quad\text{CH}$$
$$\quad\quad\quad\quad\;|$$
$$\quad\quad\quad\text{CH}_2\text{SH}$$

Polymers (XLIII–XLV) containing vicinal mercaptan groups were also synthesized from poly(vinyl alcohol) (56, 57). The bromoacetal was elaborated with sodium allyloxide, and the allyl ether formed was converted to a dithiol by addition of bromine followed by substitution with KSH. The incorporation of free mercaptan groups was low ($\sim 25\%$) (Scheme VIII). These materials are polymeric analogs of 2,3-dithiol-1-propanol (BAL) and would probably be very efficient heavy metal scavengers.

Scheme VIII

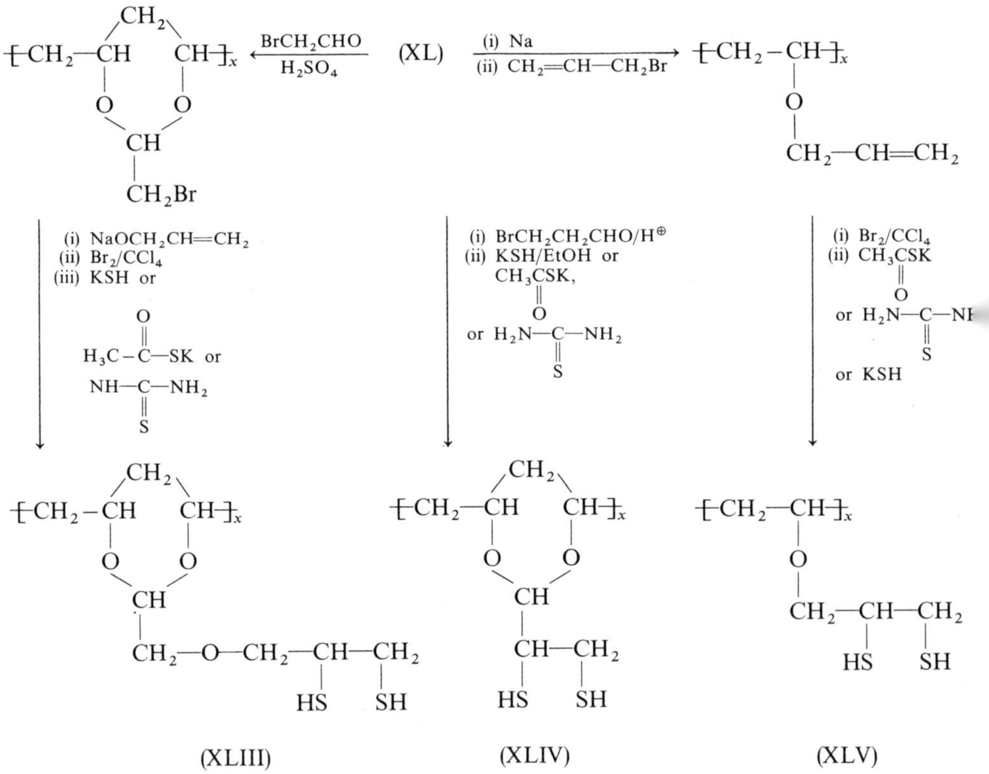

c) Polystyrene Derivatives

Chemical introduction of mercaptan groups into polystyrene derivatives has been extensively examined (Scheme IX). Poly(p-mercaptostyrene) (XLVI) was prepared by Gregor et al. (2) by treating

Scheme IX

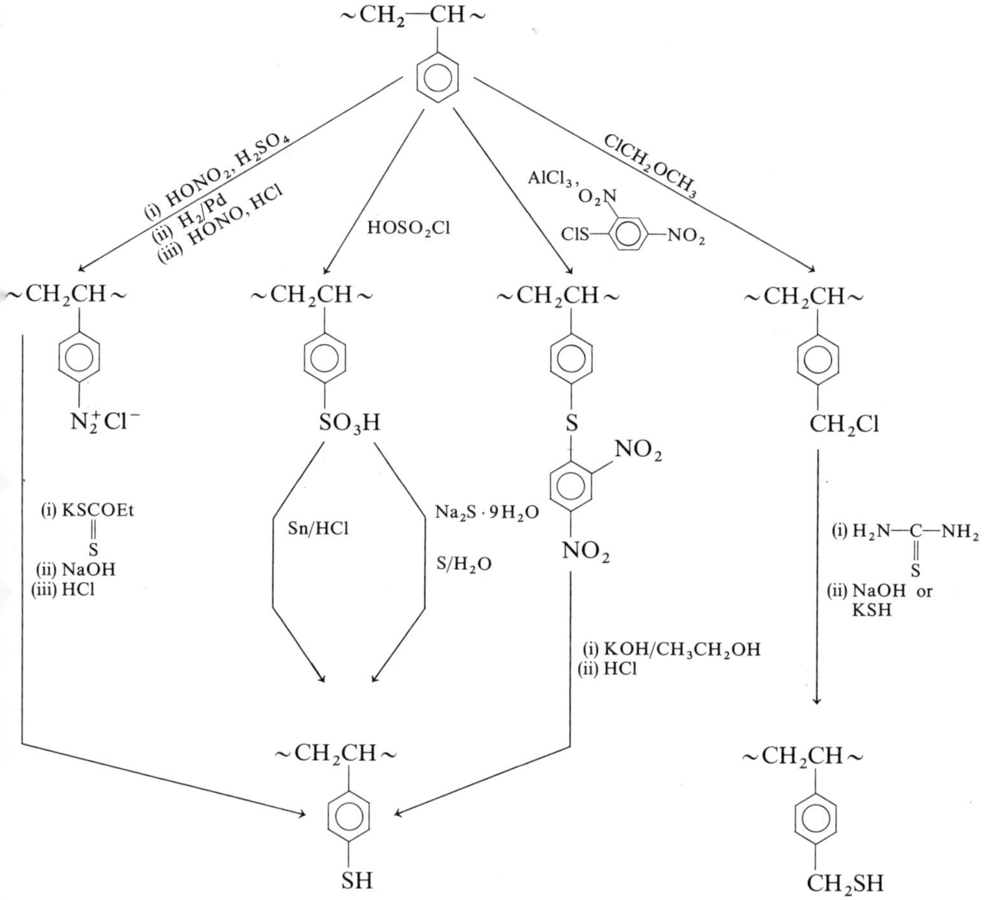

diazotized poly(p-amino styrene) with potassium ethyl xanthate; following alkaline hydrolysis a polymer assaying at 90% of the theoretical sulfur content was obtained. The insoluble polymer was swollen by alkali and was effective as an ion exchange resin. Poly(p-mercaptostyrene) with 15% of the theoretical sulfur content was made by reducing crosslinked chlorosulfonated polystyrene (58–60).

Alternatively, (XLVI) was prepared by treating polystyrene with 2,4-dinitrophenylsulfenyl chloride followed by alcoholysis. The extent of incorporation of free mercaptan groups was low ($\sim 20\%$) (58).

Poly(p-lithiostyrene), a highly reactive polymer prepared from iodinated polystyrene and lithium, yielded a copolymer containing 62 mol-% p-mercaptostyrene when sulfur was added (61). Similarly, treatment of styrene-divinylbenzene copolymers with sulfur and aluminum chloride yielded sulfur-containing polymers in which 21% of the incorporated sulfur is present in the form of mercaptan groups (62).

d) Chloromethylated Polystyrene Derivatives

Poly(vinylbenzyl mercaptan) has been described by several authors. The synthesis always follows the same path: chloromethylation of polystyrene followed by reaction with thiourea and hydrolysis of the isothiuronium salt (1, 53, 55, 63–65). The redox properties of a polymer obtained from chloromethylated styrene-divinylbenzene copolymers have been evaluated (3). Redox capacities given for mercaptyl resins were determined to be 2.40–5.27 milliequivalents of iodine reduced per gram of (dry) resin in aqueous potassium iodide. The oxidized form of the resin could be easily reduced with 10% aqueous bisulfite for a complete redox cycle. Recently polyvinylbenzyl mercaptan resins were prepared directly by treating chloromethylated polystyrene with KSH in dimethyl formamide (66).

Scheme X

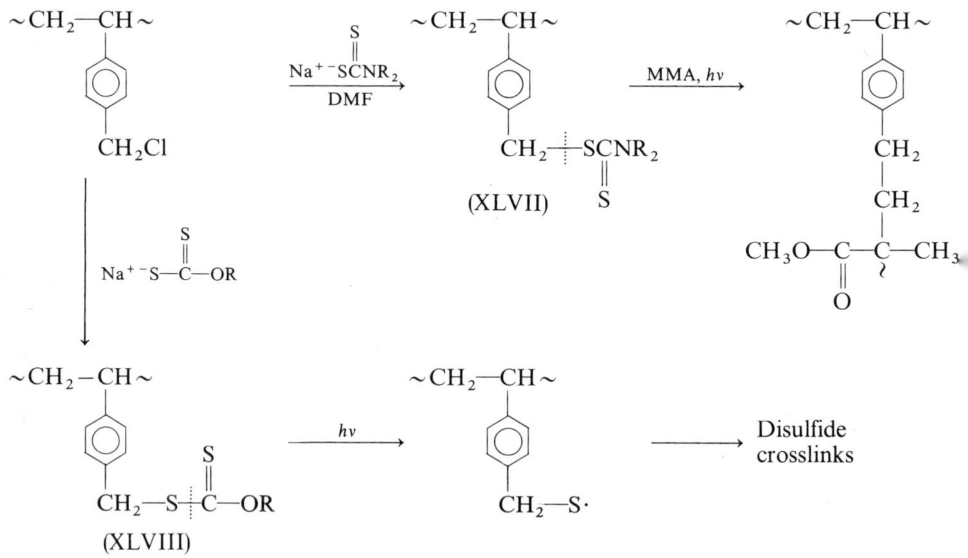

Chloromethylated polystyrene has also been used to synthesize polymers containing vicinal mercaptan (*57*), dithiocarbamate (XLVII), and xanthate groups (*43,68*) (XLVIII). The photochemical reactions of polymers (XLVII, XLVIII) and their model compounds have been extensively studied by Okawara et al. (*67,70*). Photolysis of polymers containing dithiocarbamate groups (XLVII) in the presence of vinyl monomers such as methyl methacrylate (MMA) or styrene, generated graft copolymers (*43,69*), which did not contain sulfur (Scheme X). In contrast, polymers containing xanthate groups (XLVIII) produced polyvinylbenzyl thiol radicals upon irradiation (*43,68*) and crosslinked *via* disulfide linkages. The potential applications of these polymers as photoresist resins were considered.

The introduction of N,N-diethyldithiocarbamate and N-methyldithiocarbamate groups into poly(vinyl chloride) has been reported (*11,12*). Thermal decomposition of polymers containing N-methyldithiocarbamate groups (IL) failed to produce mercaptan groups, but formed a cyclic structure on the polymer matrix (Scheme XI). Pyrolysis of polymer containing N,N-diethyldithiocarbamate (L) produced polyenes. However, the polymers reacted with nucleophiles such as dimethylamine in dioxane to yield insoluble polymercaptans. LiAlH$_4$ reduction of (L) failed to produce a polymeric thiol. It was also reported that poly(vinyl mercaptan) was prepared from PVC and KSH, but this reaction was accompanied by extensive dehydrohalogenation (*71*).

Scheme XI

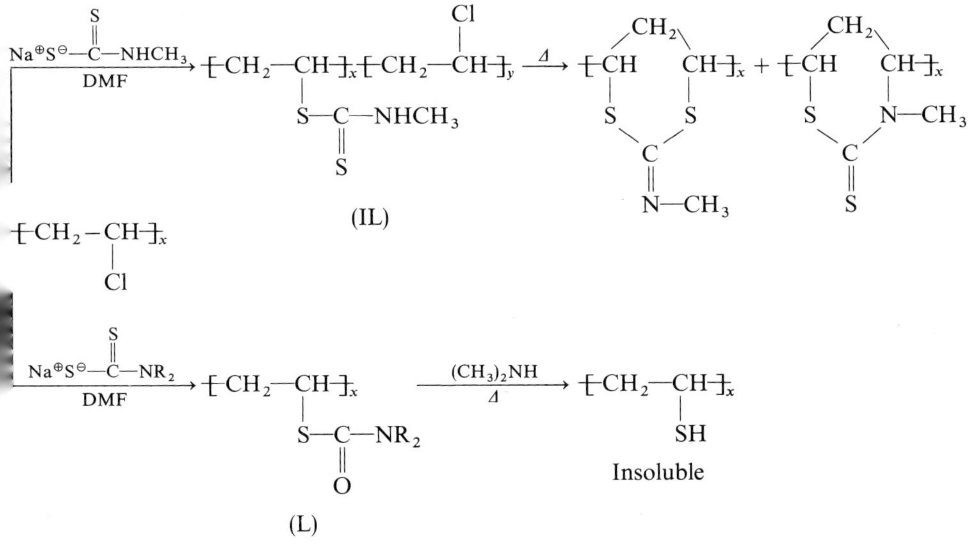

e) Reactions With Bender's Salts

The application of the nucleophilic properties of Bender's salts to the displacement of reactive halides on polymer substrates has been thoroughly evaluated by Daly and Lee (72). Polymers containing halomethyl functional groups such as: chloromethylated polystyrene, poly(epichlorohydrin), copoly(epichlorohydrin-ethylene oxide), poly(2,6-bis-bromomethylphenylene oxide) and poly(vinyl chloride) were utilized as substrates with varying degrees of success.

Relatively rigid polymeric substrates containing isolated reactive sites reacted quantitatively with Bender's salts to produce the desired polymeric thiocarbonates (LI). However, quantitative conversion of the functional groups of flexible polymers was not achieved due to the propensity for S-alkyl-O-t-alkylthiocarbonates to undergo side reactions leading to crosslinked polymers (LIII) which can not be converted to polymercaptans (Scheme XII).

Scheme XII

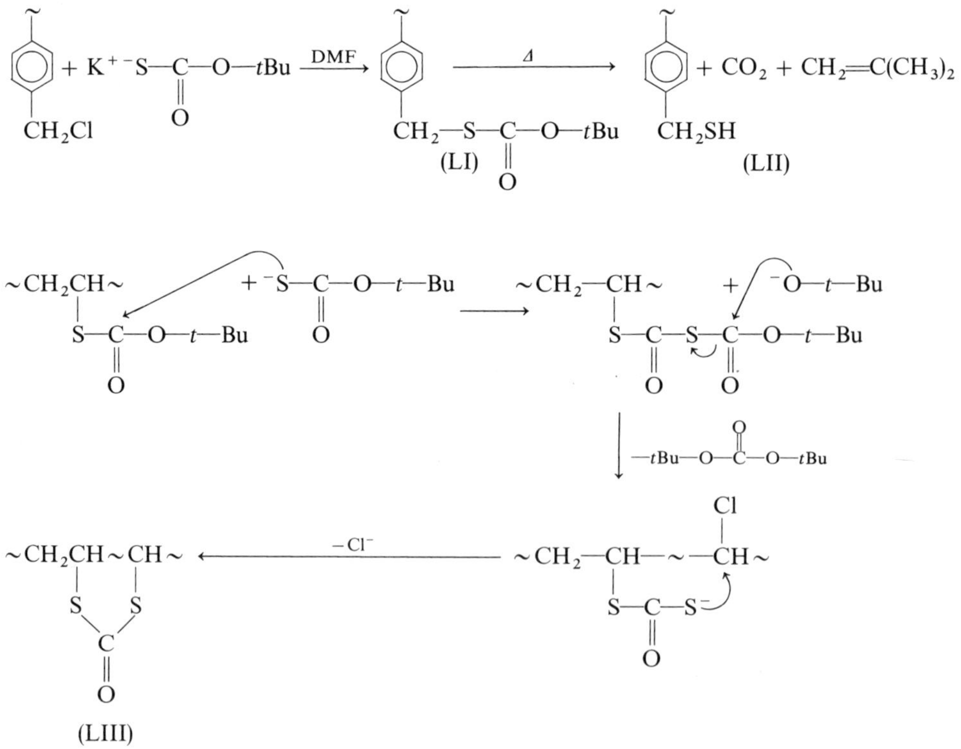

$$\text{Ar-CH}_2\text{Cl} + \text{K}^{+\,-}\text{S-C(=O)-O-}t\text{Bu} \xrightarrow{\text{DMF}} \text{Ar-CH}_2\text{-S-C(=O)-O-}t\text{Bu} \xrightarrow{\Delta} \text{Ar-CH}_2\text{SH} + CO_2 + CH_2=C(CH_3)_2$$

(LI) → (LII)

[Crosslinking side-reaction pathway leading to (LIII)]

Okawara et al., have prepared a vicinal mercaptan-containing polymer from poly(acryloyl chloride) (57) by similar procedures for preparation of XLIII. They also prepared a water-soluble dithiocarbamate-polymer containing nitrogen in the backbone chain from poly(ethylene imine) (43, 74) and a dithiocarbamate derivative from cellulose-acetate chloroacetate (43). Their photochemical behavior and ion-exchange capacity were similar to that exhibited by the poly(vinylbenzyl)-derivatives. These polymers were also used as efficient heavy metal scavengers.

Gluckman et al. (75) reacted thioglycolic acid with copolymers containing the glycidyl methacrylate residue to produce resins containing two pendant mercaptan groups for each epoxide group.

$$\{CH_2-CH\}_x \quad + \quad 2\,HSCH_2COH \quad \longrightarrow \quad \{CH_2-CH\}_x$$

(with CH$_3$ and C(=O)OCH$_2$CH—CH$_2$ epoxide side group on left; product on right bearing COCH$_2$CHO—C(=O)—CH$_2$SH and CH$_2$O—C(=O)—CH$_2$SH pendant groups)

Ion-exchange resins derived from chloromethyl acrylate and thioglycolic acid (75) and a mercaptan-containing polymer from copolymer of vinyl acetate with 1,3-dichloro-2-butene (76) have also been reported.

VI. Introduction of Mercapto Groups into Proteins

Several successful attempts have been made at introducing mercapto groups into proteins without degrading the polypeptide chain. N-acetylhomocysteine thiolactone (LIV) in the presence of silver ion reacted readily with the primary amino group of lysine residues to produce thiolated proteins (LV) (77).

$$\text{(LIV)} \quad \underset{\substack{CH_2 \\ | \\ CH_2-CH-NH-\underset{\underset{O}{\|}}{C}-CH_3}}{\overset{S}{\diagup}\overset{\diagdown}{\underset{|}{C=O}}} + NH_2\text{-Protein} \longrightarrow \underset{\text{(LV)}}{HSCH_2CH_2\underset{\substack{| \\ NH \\ | \\ C=O \\ | \\ CH_3}}{CH}\overset{O}{\overset{\|}{C}}-NH\text{-Protein}}$$

An alternate method used thioglycolides, to prepare highly thiolated casein and ovalbumin (78).

$$+S-CH_2-\overset{O}{\overset{\|}{C}}-S-CH_2-\overset{O}{\overset{\|}{C}}\!\!+_x + \overset{H_2N}{\underset{H_2N}{\diagup}}\!\!\text{Protein} \longrightarrow \overset{HSCH_2\overset{O}{\overset{\|}{C}}-NH}{\underset{HSCH_2\underset{\underset{O}{\|}}{C}-NH}{\diagup}}\!\!\text{Protein}$$

Still another method used S-acetylmercaptosuccinic anhydride (6).

$$\underset{\substack{| \\ CH_2-C \\ \diagdown \\ O}}{CH_3\overset{O}{\overset{\|}{C}}-S-CH-\overset{\diagup O}{C\diagdown_O}} + H_2N\text{—Protein} \longrightarrow \underset{CH_2-COOH}{CH_3\overset{O}{\overset{\|}{C}}S-CH-\overset{O}{\overset{\|}{C}}-NH\text{—Protein}}$$

$$\downarrow NaOH$$

$$\underset{CH_2-COOH}{HS-CH-\overset{O}{\overset{\|}{C}}-NH\text{—Protein}}$$

The excess thiol functions are useful for binding enzymes to polymer-captan supports using reversible disulfide linkages.

Recently, Barker, and Gray (79) have devised a new type of acrylamide enzyme support system which utilizes mercaptan groups as the active functional sites. The carrier, Enzacryl Polythiol (LVI), was derived from a copolymer of acrylamide, N-acryloyl-S-benzylcysteine and N,N'-methylene-*bis*-acrylamide. Free mercaptan groups were exposed by removal of the benzyl blocking groups with sodium in liquid ammonia. Enzycryl Polythiol couples oxidatively with thiolated enzymes (LV) to form disulfide linked bound enzymes (LVII) (Scheme XIII).

Scheme XIII

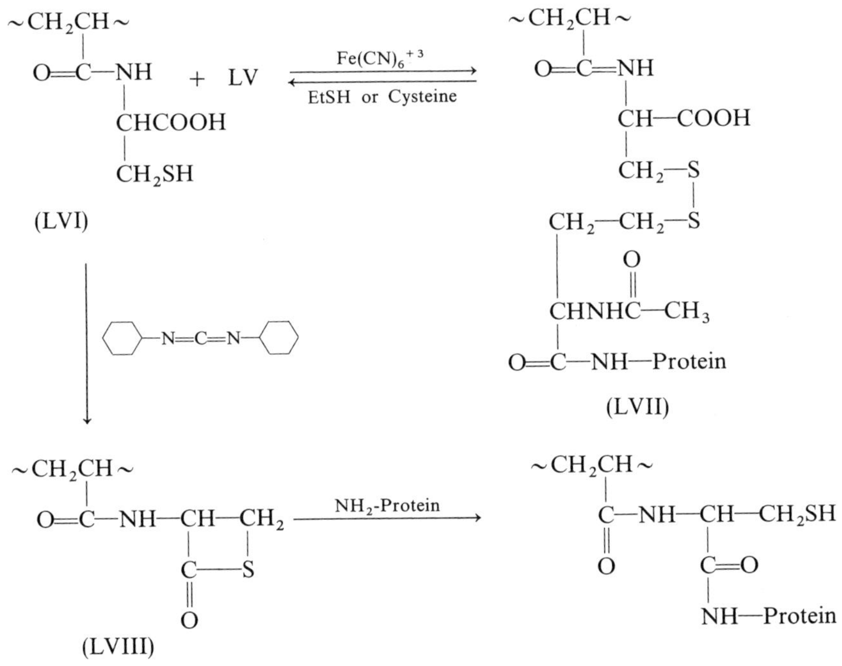

$$\sim CH_2CH \sim$$
$$|$$
$$O=C-NH \quad + \quad LV \quad \xrightleftharpoons[\text{EtSH or Cysteine}]{Fe(CN)_6^{+3}} \quad \begin{array}{c} \sim CH_2CH \sim \\ | \\ O=C-NH \\ | \\ CH-COOH \\ | \\ CH_2-S \\ | \\ CH_2-CH_2-S \\ | \\ O \\ \| \\ CHNHC-CH_3 \\ | \\ O=C-NH-\text{Protein} \end{array}$$

(LVI) → (via dicyclohexylcarbodiimide) → (LVIII)

(LVII)

$$\begin{array}{c} \sim CH_2CH \sim \\ | \\ O=C-NH-CH-CH_2 \\ | \quad | \\ C\!\!-\!\!-\!\!S \\ \| \\ O \end{array} \quad \xrightarrow{NH_2\text{-Protein}} \quad \begin{array}{c} \sim CH_2CH \sim \\ | \\ C-NH-CH-CH_2SH \\ \| \quad | \\ O \quad C=O \\ | \\ NH-\text{Protein} \end{array}$$

(LVIII)

A novel feature of conjugates based on Enzacryl Polythiol (LVI) is that the coupling can be reversed by treatment with cysteine or mercaptoethanol. Regenerate enzyme carriers have obvious commercial potential. Another feature of LVI is that it may also be activated for enzyme coupling with dicyclohexylcarbodiimide. The activated copolymer, Enzycryl Polythiolactone (LVIII), may be stored indefinitely under dry conditions (80). Enzyme coupling occurs *via* nucleophilic opening of the thiolactone ring; thus, the conjugate contains free mercaptan groups which may be used to further modify the bound enzyme.

VII. References

1. Parrish, J. R.: Chemistry and Industry 137 (1959).
2. Gregor, H. P., Dolar, D., Hoeshele, G. K.: J. Am. Chem. Soc. **77**, 3675 (1955).
3. Cassidy, H. G.: J. Am. Chem. Soc. **71**, 402 (1949); Kun, K. A.: J. Polymer Sci., Part A-1, **4**, 847 (1966).
4. — Ezrin, E., Updegraff, I. H.: J. Am. Chem. Soc. **75**, 1610, 1615 (1953).
5. Barron, E. S. G.: Adv. Enzymology, Vol. XI, pp. 219. New York: Interscience Publ. Inc. 1951.
6. Overberger, C. G., Ferraro, J. J., Bonsignore, P. V., Orttung, F. W., Vorchheimer, N.: Pure Appl. Chem. **4**, 521 (1962).
7. Bacq, Z. M.: Chemical Protection against Ionizing Radiation, Thomas, Springfield, Ill., 1965.
8. Moenig, H., Ringsdorf, H.: Makromol. Chem. **127**, 204 (1969).
9. Gunnung, H. E.: U.S. Patent 3474016 (Oct., 1969); Chem. Abstr. **72**, 54733w (1970).
10. Straus, O. P., Hikida, T., Gunning, H. E.: Can. J. Chem. **43**, 717 (1965).
11. Okawara, M., Marishita, K., Imoto, E.: Kogyo Kagaku Zasshi **69**, 761 (1966).
12. Nakagawa, T., Taniguchi, Y., Okawara, M.: Kogyo Kagaku Zasshi **70**, 2382 (1967).
13. (a) Kern, W., Schulz, R. C.: Angew. Chem. **69**, 153 (1957); (b) Cohen, H. L., Minsk, L. M.: J. Org. Chem. **24**, 1404 (1959).
14. Kun, K. A., Cassidy, H. G.: J. Polymer Sci. **44**, 383 (1960).
15. Brubaker, M. M.: U.S. Pat. 2378535 and 2378536 (1945); Chem. Abstr. **39**, 4331 (1945).
16. Overberger, C. G., Biletch, H., Nickerson, R. G.: J. Polymer Sci. **27**, 381 (1958).
17. — Ferraro, J. J., Orttung, F. W.: J. Org. Chem. **26**, 3458 (1961).
18. Hardy, G., Vargu, J., Nyrtrai, K., Czajlik, I., Zubonyai, L.: Vysokomolekul. Soedin. **6**, 758 (1964).
19. Imoto, M., Kinoshita, M., Irie, T.: Kogyo Kagaku Zasshi **72**, 1210 (1969).
20. Overberger, C. G., Ferraro, J. J.: J. Org. Chem. **27**, 3539 (1962).
21. Richards, L. M. (DuPont): U.S. Patent 2418426 (1947); Chem. Abstr. **41** 4011d (1947).
22. Leonard, F., Hanks, G. A., Kulkarni, R. K.: J. Polymer Sci., Part A-1, **4**, 449 (1966).
23. Nakazawa, S., Kinoshita, M., Imoto, M.: Kogyo Kagaku Zasshi **70**, 1452 (1967).
24. Overberger, C. G., Lebovits, A.: J. Am. Chem. Soc. **78**, 4792 (1956).
25. — — J. Am. Chem. Soc. **77**, 3675 (1955).
26. Nummy, W. R. (to Dow Chem. Co.): U.S. Patent 2947731 (1960); Chem. Abstr. **31**, 4045a (1961).
27. Nelson, S. J., Jr. (to United States Rubber Co.): U.S. Patent 3260748 (1966); Chem. Abstr. **65**, 10528c (1966).
28. (a) Gollmer, K., Ringsdorf, H.: Makromol. Chem. **121**, 227 (1969); (b) Kroker, R., Ringsdorf, H.: Makromol. Chem. **121**, 240 (1969).
29. Gollmer, K., Muller, F. H., Ringsdorf, H.: Makromol. Chem. **92**, 122 (1966).
30. Gollmer, K., Ringsdorf, H.: Kolloid-Z. u. Z. Polymere **216**, **217**, 325 (1967).
31. Ringsdorf, H., Overberger, C. G.: J. Polymer. Sci. **61**, S11 (1962).
32. Matsoyan, S. G., Soakyan, A. A.: Izv. Akad. Nauk. Arm. SSR, Khim. Nauki **15** (5), 463 (1962); Chem. Abstr. **59**, 7655 (1963); Vysokomolekul. Soedin. **3**, 1755 (1961); Chem. Abstr. **56**, 14443g (1962).

33. (a) Corey, E. J., Seebach, D.: Angew. Chem. I.E. **4**, 1075, 1077 (1965); (b) Seebach, D.: Synthesis **1**, 17 (1969).
34. Ringsdorf, H., Overberger, C. G.: Makromol. Chem. **44-46**, 418 (1961).
35. Overberger, C. G., Daly, W. H.: J. Org. Chem. **29**, 757 (1964).
36. Daly, W. H., Ph. D. Dissertation, Polytechnic Institute of Brooklyn, 1965.
37. Overberger, C. G., Ringsdorf, H., Avchen, B.: J. Org. Chem. **30**, 3088 (1965).
38. — — Weinshenker, N.: J. Org. Chem. **27**, 4331 (1962); Makromol. Chem. **64**, 126 (1963).
39. — Daly, W. H.: J. Am. Chem. Soc. **86**, 3402 (1964).
40. — Burg, K. H., Daly, W. H.: J. Am. Chem. Soc. **87**, 125 (1965).
41. Daly, W. H., Lee, Chien-Da S., Overberger, C. G.: J. Polymer Sci., Part A-1, **9**, 1723 (1971).
42. Nakai, T., Shioya, K., Okawara, M.: Makromol. Chem. **108**, 95 (1967).
43. Okawara, M., Nakai, T.: Bull. Tokyo Inst. Technol. **78**, 1 (1966); Chem. Abstr. **68**, 7873ln (1968).
44. Potassium O-ethylthiocarbonate was originally synthesized by C. Bender in 1868, and the entire class of compounds with the general formula,

$$\text{R---O---}\underset{\|}{\overset{O}{C}}\text{---S}^{\ominus \oplus}\text{M,}$$

have been named in his honor. See, e.g. Bender, C.: Liebigs Ann. Chem. **148**, 137 (1868); Bull. Soc. Chim. (2), **12**, 256 (1869); Fischer, R., Fessler, G.: Pharmazie **10**, 349 (1955).
45. Dykstra, T. K., Smith, D. A.: U.S. Patent 3306844, Feb., 1967; Chem. Abstr. **66**, 116153u; Makromol. Chem. **134**, 209 (1970).
46. Shashoua, V. E.: U.S. Patent 3179638, April, 1965; Chem. Abstr. **63**, 3127b (1965).
47. Berger, A., Nogushi, J., Katchalski, E.: J. Am. Chem. Soc. **78**, 4483 (1956).
48. Ida, T., Noda, K., Takahashi, S., Utsumi, I.: Makromol. Chem. **73**, 215 (1964).
49. Goethals, E. J., Du Prez, E.: J. Polymer Sci., Part A-1, **4**, 2893 (1966); Makromol. Chem. **146**, 145 (1971).
50. Overberger, C. G., Aschkenasy, H.: J. Org. Chem. **25**, 1648 (1960); J. Am. Chem. Soc. **82**, 4357 (1960).
51. Hamamura, Y., Tatsukawa, M., Uno, S.: Nippon Nogei-Kagaku Kaishi **29**, 194 (1955); Chem. Abstr. **52**, 20772d (1958).
52. Cairns, T. L., Gray, H. W., Schneider, A. K., Schreiber, R. S.: J. Am. Chem. Soc. **71**, 655 (1949).
53. Okawara, M., Nakagawa, T., Imoto, E.: Kogyo Kagaku Zasshi **60**, 73 (1957); Chem. Abstr. **53**, 5730d (1959).
54. Nakamura, Y.: Kogyo Kagaku Zasshi **58**, 269 (1955); Chem. Abstr. **49**, 14376h (1955).
55. Cerny, J., Wichterle, O.: J. Polymer Sci. **30**, 501 (1958).
56. Okawara, M., Sumitomo, Y.: Bull. Univ. Osaka Perfect. Ser. A **6**, 119 (1958); Chem. Abstr. **53**, 7003c (1959); Kogyo Kagaku Zasshi **61**, 1508 (1958); Chem. Abstr. **56**, 1330i (1962).
57. — Haruki, E., Imoto, B.: Kogyo Kagaku Zasshi **64**, 229 (1961); Chem. Abstr. **57**, 4853f (1962).
58. — Onishi, Y., Imoto, E.: Kogyo Kagaku Zasshi **64**, 226 (1961); Chem. Abstr. **57**, 4853d (1962).
59. Seifert, H.: Ger. Pat. No. 1067217 (Oct., 1959); Chem. Abstr. **55**, 1075a (1961).
60. Davankov, A. B., Zambrovskaya, E. V., Vakova, I. N.: U.S.S.R. Pat. 146941 (1962); Chem. Abstr. **57**, 11408b (1962).

61. Braun, D.: Makromol. Chem. **30**, 85 (1959); Chimia **14**, 24 (1960).
62. Davankov, A. B., Zambrovskaya, E. V., Gerashchenko, Z. V.: Vysokomolekul. Soedin. **3**, 1468 (1961).
63. Arcus, C. L., Solomons, N. S.: J. Chem. Soc. 1175 (1963).
64. Trostyanskaya, E. B., Tevlina, A. S.: Zhur. Anal. Khim. **15**, 402 (1960); Chem. Abstr. **55**, 12703a (1961).
65. Davankov, A. B., Zambrovskaya, E. V.: Vysokomolekul. Soedin. **2**, 1330 (1960).
66. Crowley, J. I., Harvey, III, T. B., Rapoport, H.: J. Macromol. Sci. Chem. A **7**, 1117 (1973).
67. Okawara, M., Yamashina, H., Ishiyama, K., Imoto, E.: Kogyo Kagaku Zasshi **66**, 1383 (1963); Chem. Abstr. **60**, 14628e (1964).
68. — Nakai, T., Imoto, E.: Kogyo Kagaku Zasshi **68**, 582 (1965); Chem. Abstr. **63**, 10070f (1965).
69. — — Morishita, K., Imoto, E.: Kogyo Kagaku Zasshi **67**, 2108 (1964); Chem. Abstr. **62**, 14827a (1965).
70. — — Imoto, E.: Kogyo Kagaku Zasshi **69**, 973 (1966); Chem. Abstr. **65** 20236c (1966).
71. Hamamura, Y., Uejima, H., Ishikawa, N., Iguchi, S., Hayashiya, K.: Nippon Nogei-Kagaku Kaishi **31**, 703 (1957); Chem. Abstr. **52**, 12269d (1968).
72. Daly, W. H., Lee, Chien-Da S.: Polymer Preprints, Vol. **14**, No. 2, 1238 (1973).
73. Gluckman, M. S., Kampf, M. J., O'Brien, J. L., Fox, T. G., Graham, R. K.: J. Polym. Sci. **37**, 411 (1959).
74. Okawara, M., Ori, M., Nakai, T., Imoto, E.: Kogyo Kagaku Zasshi **69**, 766 (1966); Chem. Abstr. **65**, 20236b (1966).
75. Rohm and Haas, G.m.b.H. (by H. Zima): German Patent 1 041 250 (1958); Chem. Abstr. **54**, 20015d (1960).
76. Durgaryan, A. A., Grigoryan, A. S., Chaltykyan, O. A.: Izv. Akad. Nauk. Arm. SSR., Khim. Nauki **15** 455 (1962); Chem. Abstr. **58**, 14104g (1963).
77. Benesch, R., Benesch, R. E.: In: Sulfur in Proteins. Academic Press, New York (1959), pp. 15ff.; J. Am. Chem. Soc. **78**, 1597 (1956).
78. Schoberl, A.: Angew. Chem. **60**, 7 (1948).
79. Barker, S. A., Gray, C. J., Khoujah, A. M., Law, C. M. J.: Birmingham University Chemical Engineering Journal (1971).
80. Epton, R., Thomas, T. H.: Aldrichimica Acta **4**, 61 (1971).

Received January 17, 1974

Structures of Copolymers of High Olefins

YU. V. KISSIN

Institute of Chemical Physics
Academy of Sciences of the USSR, Moscow, USSR

Table of Contents

I. Introduction	92
II. Statistical Description of Binary Copolymers	92
A. The Enantiomorphous Model	93
B. The Racemic Model	96
III. Experimental Methods for the Study of Copolymer Structures	98
A. Synthesis of Copolymers in Stationary Conditions	98
B. Calculations of r_1 and r_2	98
C. IR Methods	99
D. Nuclear Magnetic Resonance	106
E. Melting-point Measurements	107
F. Crystallinity Measurements	108
IV. Structure of High-olefin Copolymers	109
A. Ethylene Copolymers	110
B. Propylene Copolymers	121
C. Butene-1 Copolymers	128
D. 3-Methylbutene-1 Copolymers	131
E. 4-Methylpentene-1 Copolymers	133
F. Styrene Copolymers	138
G. Optically Active Copolymers	140
V. Compositional Inhomogeneity of Olefin Copolymers	141
A. r_1 and r_2 Values in Multiple-site Model	142
B. Structure of Copolymer Fractions	143
VI. Activities of Different High Olefins in Homopolymerization and Copolymerization Reactions	147
A. Reactivity Ratios for High Olefin Copolymerization	147
B. Relative Activities of Olefins in Homopolymerization and Copolymerization Reactions	147
VII. Conclusion	149
VIII. Acknowledgements	150
IX. References	151

I. Introduction

The discovery of the transition-metal catalysts for olefin polymerization made possible the synthesis not only of various stereoregular olefin polymers but also that of olefin copolymers.

The main subject of this review is the structure of high-olefin copolymers. The copolymers of high olefins are binary copolymers in which one or both of the comonomers are olefins containing more than three carbon atoms. Thus, we include in the review the copolymers of such olefins as butene-1 and the higher linear olefins, and branched olefins like 3-methylbutene-1 or 4-methylpentene-1, also the copolymers of styrene and some other aryl monomers obtained with complex catalysts.

The ethylene–propylene copolymers are mentioned briefly and only for the sake of comparison with other copolymers. Among the problems discussed are the determination of composition, the chemical structure of the copolymers, the qualitative and quantitative evaluation of the distribution of monomer units in the copolymers, and the compositional nonuniformity of olefin copolymers. The review also covers two related questions: the estimation of r_1 and r_2 values for high-olefin copolymers, and the comparison of the activities of different olefins in homopolymerization and copolymerization reactions.

The scope of the discussions explains the choice of the experimental methods covered in the paper, mainly IR spectroscopy, X-ray methods, melting-point measurements, NMR spectra, etc.

II. Statistical Description of Binary Copolymers

The statistics of binary copolymers is one of the most advanced parts of polymer kinetics. All the necessary equations for the copolymer composition and monomer distribution in the simple four-reaction scheme (1, 2), for the penultimate effect (3, 4) and the pen-penultimate effect (5) wer obtained by the kinetic method and by an approach based on the Markov chain theory (6). Some of the recent papers in this field are very sophisticated (5, 7, 8) and the theoretical achievements often far exceed the experimental possibilities for testing them.

Structures of Copolymers of High Olefins 93

When examining binary copolymers obtained with stereospecific catalysts, it is important to consider not only the effects of mixing the monomer units in the chain but also the stereoregulating effect of the catalytic system on the structure of the sequences.

If we neglect penultimate and higher-order effects in the copolymerization of monomers M_1 and M_2 and assume that we have a standard scheme of chain growth with the complex catalytic system, the following set of reactions takes place:

1. $\sim D_1^* \ldots \text{Cat} + M_1 - K_{D_1 D_1} \rightarrow \sim D_1 D_1^* \ldots \text{Cat}$
2. $\sim D_1^* \ldots \text{Cat} + M_1 - K_{D_1 L_1} \rightarrow \sim D_1 L_1^* \ldots \text{Cat}$
3. $\sim L_1^* \ldots \text{Cat} + M_1 - K_{L_1 L_1} \rightarrow \sim L_1 L_1^* \ldots \text{Cat}$
4. $\sim L_1^* \ldots \text{Cat} + M_1 - K_{L_1 D_1} \rightarrow \sim L_1 D_1^* \ldots \text{Cat}$
5. $\sim D_1^* \ldots \text{Cat} + M_2 - K_{D_1 D_2} \rightarrow \sim D_1 D_2^* \ldots \text{Cat}$
6. $\sim D_1^* \ldots \text{Cat} + M_2 - K_{D_1 L_2} \rightarrow \sim D_1 L_2^* \ldots \text{Cat}$
7. $\sim L_1^* \ldots \text{Cat} + M_2 - K_{L_1 L_2} \rightarrow \sim L_1 L_2^* \ldots \text{Cat}$
8. $\sim L_1^* \ldots \text{Cat} + M_2 - K_{L_1 D_2} \rightarrow \sim L_1 D_2^* \ldots \text{Cat}$
9. $\sim D_2^* \ldots \text{Cat} + M_2 - K_{D_2 D_2} \rightarrow \sim D_2 D_2^* \ldots \text{Cat}$
10. $\sim D_2^* \ldots \text{Cat} + M_2 - K_{D_2 L_2} \rightarrow \sim D_2 L_2^* \ldots \text{Cat}$
11. $\sim L_2^* \ldots \text{Cat} + M_2 - K_{L_2 L_2} \rightarrow \sim L_2 L_2^* \ldots \text{Cat}$
12. $\sim L_2^* \ldots \text{Cat} + M_2 - K_{L_2 D_2} \rightarrow \sim L_2 D_2^* \ldots \text{Cat}$
13. $\sim D_2^* \ldots \text{Cat} + M_1 - K_{D_2 D_1} \rightarrow \sim D_2 D_1^* \ldots \text{Cat}$
14. $\sim D_2^* \ldots \text{Cat} + M_1 - K_{D_2 L_1} \rightarrow \sim D_2 L_1^* \ldots \text{Cat}$
15. $\sim L_2^* \ldots \text{Cat} + M_1 - K_{L_2 L_1} \rightarrow \sim L_2 L_1^* \ldots \text{Cat}$
16. $\sim L_2^* \ldots \text{Cat} + M_1 - K_{L_2 D_1} \rightarrow \sim L_2 D_1^* \ldots \text{Cat}$.

Here "Cat" means a catalytic site, D_1 and L_1 stand for the stereoisomeric units of the first monomer in the chain, and D_2 and L_2 for the second monomer. Reactions 1–4 correspond to the stereospecific polymerization of the monomer M_1, and reactions 9–12 to that of the monomer M_2.

The following two cases of particular importance were considered (9, 10).

A. The Enantiomorphous Model

In examining the stereospecific polymerization on heterogeneous catalysts, it can be assumed in the first approximation that the polymer regularity is determined mainly by the geometry of the active site ["Cat" in reactions (*1–16*)] rather than by the structure of the last monomer unit in the chain, *i.e.*

$$K_{D_1D_1} = K_{L_1D_1}, \quad K_{L_1L_1} = K_{D_1L_1}, \quad K_{D_2D_2} = K_{L_2D_2}, \quad \text{and} \quad K_{L_2L_2} = K_{D_2L_2}.$$

Two alternative types of sites exist on the catalyst surface in equal numbers: D sites yielding predominantly D-isotactic polymer chains of both M_1 and M_2 (like DDDDDLDDDD), and L sites yielding predominantly L-isotactic chains (*11*). For a D site, $K_{D_iD_i} = K_{L_iD_i} > K_{L_iL_i} = K_{D_iL_i}$; for an L site the opposite takes place. This model corresponds to the idea of matrix catalysis on a stereospecific active site. Its validity has been proved by certain spectral data (*12, 13, 133, 134*) and is supported by studies of models of the stereospecific active site (*14–16*).

Statistical expressions for the stereoregular homopolymers in the model are given in (*11, 17, 18*). The extrapolation of this model to binary copolymerization (*9*) gives the following results. The parameters of the stereospecificity of the sites (all sites are assumed to have the same specificity):

$$R' = K_{D_1D_1}/K_{D_1L_1} = K_{L_1D_1}/K_{L_1L_1};$$
$$R'' = K_{D_2D_2}/K_{D_2L_2} = K_{L_2D_2}/K_{L_2L_2}. \tag{1}$$

For D sites R' and $R'' > 1$, and for L sites R' and $R'' < 1$. The numbers of D and L sites are assumed to be equal and if their stereospecificities are the same, then the equations for a copolymer grown on D sites are also valid for the whole product.

The standard equation for copolymer composition is also valid for the scheme 1–16:

$$f = F(r_1 F + 1)/(r_2 + F). \tag{2}$$

Here $f = (M_1/M_2)_{cop}$, $F = (M_1/M_2)_{feed}$ and the reactivity ratios are

$$r_1 = K_{M_1M_1}/K_{M_1M_2} = K_{D_1L_1}(1 + R')/K_{D_1L_2}(1 + R'')$$
$$r_2 = K_{M_2M_2}/K_{M_2M_1} = K_{D_2L_2}(1 + R'')/K_{D_2L_1}(1 + R'). \tag{3}$$

The ratios between the D and L stereoisomers in a D-site copolymer are:

$$(D_1)/(L_1) = R', \quad (D_1) + (L_1) = 1; \quad (D_2)/(L_2) = R'', \quad (D_2) + (L_2) = 1. \quad (4)$$

The probabilities of the isotactic additions are:

$$p_{D_1 D_1} = r_1 R' F/(1 + R')(1 + r_1 F); \quad p_{L_1 L_1} = r_1 F/(1 + R')(1 + r_1 F). \quad (5)$$

The fractions of the monomer units M_1 in an isotactic block of length n are:

$$\delta(D_1)_n = n(r_1 R' F)^{n-1}(r_1 F + R' + 1)^2/(1 + R')^{n+1}(1 + r_1 F)^{n+1} \quad (6)$$

$$\delta(L_1)_n = n(r_1 F)^{n-1}(r_1 R' F + R' + 1)^2/(1 + R')^{n+1}(1 + r_1 F)^{n+1}$$

and

$$\delta(\text{iso})_n = (D_1) \cdot \delta(D_1)_n + (L_1) \cdot \delta(L_1)_n \quad (7)$$

(this last function is normalized to the total quantity of M_1 in a copolymer).

The fraction of the monomeric units M_1 in the sum of isotactic blocks with lengths from $n+1$ to ∞, normalized to the total quantity of M_1 in a copolymer, is

$$\varkappa(\text{iso})_{n+1} = (D_1)\left[1 - \sum_n^\infty \delta(D_1)_n\right] + (L_1)\left[1 - \sum_n^\infty \delta(L_1)_n\right] \quad (8)$$

where the δ functions are those in (6).

Similar expressions for the distribution of M_2 units can be easily obtained from Eqs. (5) to (8) by substituting r_2 for r_1, R'' for R', and $1/F$ for F.

If we neglect the distribution of monomer units in the stereosequences and examine the distribution of the units in the chemical sequences of a given monomer, we can use the standard equations for binary copolymer (*19*):

$$p_{M_1 M_1} = r_1 F / (1 + r_1 F) \tag{9}$$

$$\delta(M_1)_n = n p_{M_1 M_1}^{n-1} \cdot (1 - p_{M_1 M_1})^2 = n(r_1 F)^{n-1}/(1 + r_1 F)^{n+1}, \tag{10}$$

$$\varkappa(M_1)_{n+1} = 1 - \sum_n^\infty \delta(M_1)_n = (r_1 F)^n (r_1 F + n + 1)/(1 + r_1 F)^{n+1}. \tag{11}$$

When we take into account the Eqs. (20), (21):

$$r_1 F = 0.5 [(f - 1) + \sqrt{(f - 1)^2 + 4 r_1 r_2 f}], \tag{12}$$

which follows directly from the composition Eq. (2), it is evident that all functions of the distributions (5) to (8) depend on three parameters: the copolymer composition f, the stereospecificity of the catalytic system in respect to a given monomer R (it is reasonable to assume that R is the same for homo- and copolymerization), and the reactivity ratio product $r_1 r_2$. This conclusion is equivalent to the well-known statement that the distribution of chemical units in a copolymer chain [Eqs. (9) to (11)] depends only on the copolymer composition f and the reactivity ratio product $r_1 r_2$ (*2*).

There are two important specific cases in the enantiomorphous model
1) copolymerization of any α-olefin with symmetrically substituted olefins (primarily ethylene); in this case the distribution of the olefin sequences is described by Eqs. (5) to (8) and that of ethylene by Eqs. (9) to (11);
2) copolymerization with an ideally stereospecific system; in this case $R' \to \infty$, $R'' \to \infty$, and the Eqs. (5) to (8) are reduced to (9) to (11), *i.e.* all sequences of both monomers are sterically pure.

B. The Racemic Model

This model corresponds to the case where the homopolymer regularity is determined by the interaction between the entering and the last units in the chain. This model is of extreme importance for all main copolymerization processes (radical, free ionic), and it has been elaborated in full detail [(2), Chapter 1], (5–7, 9, 22). The application of this model to the copolymerization of olefins is limited to the soluble systems yielding syndiotactic or atactic products in the case of homopolymerization (134).

The distribution of stereoregular sequences in this model has also been studied (7, 9, 24). According to the model (9):

$$K_{D_1D_1} = K_{L_1L_1}; \quad K_{D_2D_2} = K_{L_2L_2} \quad \text{(iso-addition)}$$

$$K_{D_1L_1} = K_{L_1D_1}; \quad K_{D_2L_2} = K_{L_2D_2} \quad \text{(syndio-addition)}.$$

The stereoregularity parameters are:

$$R''' = K_{D_1D_1}/K_{D_1L_1}; \quad R'''' = K_{D_2D_2}/K_{D_2L_2}. \tag{13}$$

For purely syndiotactic systems R''' and $R'''' \to 0$; for atactic systems R''' and $R'''' = 1$.

The composition equation has the traditional form (2) also in this model, and the expression for the probability of the syndiotactic addition is

$$p_{D_1L_1} = p_{D_1D_1}/R''', \tag{14}$$

where $p_{D_1D_1}$ is the same as in Eq. (5).

The statistical equations for the distribution of monomer units in isotactic and syndiotactic sequences have been given (9). The expression for the fraction of units in the sum of syndiotactic blocks with lengths from $n+1$ to ∞ is:

$$\varkappa(\text{syndio})_{n+1} = \frac{(r_1 F)^{n-1}[(1 + R''')(1 + r_1 F) + n(r_1 R''' F + R''' + 1)]}{(1 + R''')^{n+1}(1 + r_1 F)^{n+1}}$$

$$\frac{\cdot [(r_1 F + R''' + 1)^2 + 2R'''(r_1 F)^2 - (1 + R''')^2]}{\cdot [2(1 + R''')(1 + r_1 F) - r_1 F]} \tag{15}$$

A comparison of the distribution functions for the racemic model with Eq. (12) shows that here, too, they depend on three parameters: copolymer composition f, stereospecificity R''', and $r_1 r_2$.

III. Experimental Methods for the Study of Copolymer Structures

A. Synthesis of Copolymers in Stationary Conditions

The ordinary equation for the copolymer composition (2) is in fact differential and corresponds to the instantaneous copolymer composition. The equation becomes integral only when F is constant during a given run.

The simplest way to maintain stationary conditions in normal batch copolymerization experiments is to have low yields of copolymers. If the yields are lower than 10%, we can assume in first approximation that F was kept practically constant.

There are two kinds of modified batch experiments that allow us to keep F constant. The first possibility arises in cases where one of the monomers is liquid and the other is a gas, and polymerization takes place in pure liquid comonomer under constant pressure of the gaseous comonomer. The second possibility is to pass the mixture of both monomers (gases or liquids) into the reactor at a rate equal to the rate of copolymerization (25). Such experiments have setting periods, which can be avoided by the preparation of two monomer mixtures — one to fill the reactor (with the monomer ratio F) and the second for permanent feeding (with the monomer ratio f, i.e. with a ratio corresponding to the composition of the forming copolymer) (26). The other possibility of copolymer synthesis in stationary conditions is continuous copolymerization.

It is clear that all the distribution functions given in Section II are valid only for stationary conditions.

B. Calculations of r_1 and r_2

The values of the reactivity ratio products $r_1 r_2$ are the parameters of main significance for the statistical description of the copolymer structure, which is why these parameters are so carefully evaluated.

Detailed reviews of the methods used are given in (2). We mention here the very popular graphical Fineman-Ross method (28), the simple and convenient analytical method (29), and some precise methods based on ideas of linear programming (20) and curve fitting (30, 31).

C. IR Methods

IR spectroscopy is one of the most important methods for the determination of both copolymer composition and sequence distribution. The requirements to IR bands used for these purposes are practically opposite and are therefore discussed separately.

1. Choice of Analytical Bands for the Determination of Copolymer Composition

The main *a priori* requirements of an analytical band for determining composition are that there be proportionality between the absorbance and the content of the given monomer in a copolymer and insensitivity to the distribution of units. These requirements are very desirable but hardly attainable. It is well known that practically all vibrations of a given molecular group are influenced by the neighboring groups. Thus, in practice one is obliged to follow less regid but more realistic requirements. They are:
1) The preferred bands are those of modes highly localized on small groups (side groups are best of all), such as stretching or bending modes of CH_3, $CH(CH_3)_2$, C_6H_5, etc.
2) The absorption coefficients of the analytical bands for a given unit in a copolymer should not depend on the stereoregularity of the corresponding homopolymer, *i.e.* they must be the same for isotactic and atactic fractions. The reason for this is that usually bands sensitive to the stereoregularity of a polymer are also sensitive to the monomer distribution in copolymers (see Section III.C.2.).

There are also some requirements of analytical bands from the point of view of their practical utility (proper absorption coefficients; absence of strong overlap with other bands).

The reader will find in Section IV lists of the analytical bands used for measuring the composition of the copolymers discussed.

An example of the difficulties encountered in choosing analytical bands is cited in (32) for the case of the $1380\,cm^{-1}$ band in the polypropylene spectra.

All IR methods for measuring copolymer compositions are relative and need calibration with suitable standards. Three calibration methods are usually used: radiochemical standards, calibration with the homopolymer mixtures, and calibration with model compounds (33–36). Calibration with radiochemical standards is very precise and accurate but is limited by the availability of labelled olefins. Calibration with homopolymer mixtures is also very popular (33, 37–41, 43). It is a very simple method but has many drawbacks, the main one being the constitutional difference between such mixtures and real copolymers (32). Nevertheless, its use is justified if the chosen analytical bands are highly localized and if the copolymers examined have a block structure ($r_1 r_2 \gg 1$) similar to that of the polymer mixtures. The third calibration method consists in using model compounds for the determination of absorption coefficients. These model compounds are either homopolymers (42, 44–47), in which case the method is in principle the same as calibration with polymer mixtures, or special compounds with structures resembling those of characteristic groups in copolymers (13, 49, 51).

A general review of possible shapes of the calibration curves has been made (52) and some complications have been discussed (27, 53).

2. Sequence-sensitive IR Bands

The problem of the choice of bands sensitive to the distribution of monomer units in copolymers is closely connected with the problem of the regularity bands in the IR spectra of stereoregular polymers, and the problem of the vibrational spectroscopy of nonregular polymers in general [see recent reviews (54–56)].

a) Theoretical Approach. In the vibrational spectra of ideally regular polymers only those modes absorb in which the phase differences between the equivalent motions of neighboring units in the chain are either zero or equal to $2\pi k/l$, where k is the number of turns and l is the number of units in the identity period.

There are three general types of disorder of the ideal polymer chain:
1) conformational disorder in a melt or in solutions;
2) configurational disorder, *i.e.* the nonideal stereoregularity of the polymer chain, which depends on the catalyst specificity only;
3) chemical disorder due to copolymerization or isomerization of a homopolymer chain.

The spectral changes due to all these types of disorder are roughly the same. In every case the shortening of sterically regular blocks and the appearance of defects between them induce certain changes in the IR spectra (54, 55)

1) New bands arise due to defect vibrations (local modes); the position of these bands is determined by the nature of the defect. In copolymer spectra the new bands are generally speaking those of the second comonomer.
2) The strictness of selection rules is weakened and some forbidden bands can appear, but with low intensities (57).
3) According to the theory there exists dependence between the length of the regular block and the frequency of the bands (as well as their intensity) (57):

$$v_n^2 = \sum_{k=0}^{M} A_k \cos k\pi/(n+1),$$

where v_n is the frequency of the block with the length n, A_k is a constant, and M is usually 1 or 2 (58).

The dependence of v_n on the vibrational parameters of defects was studied (57, 59, 60) and was demonstrated to be sometimes rather small. For this reason, the appearance of a set of short blocks in nonregular polymers sometimes brings about the asymmetric widening of some bands and induces significant changes in their intensities. The problem of the variation of the absorption coefficients K_n with n is only beginning to be examined theoretically (61) and the main results obtained are empirical.

b) Empirical Approach. The empirical classification of the bands sensitive to chain regularity has been discussed (62). There are three kinds of IR bands suitable for the study of unit distribution (see also Fig. 1):

1) Crystallinity bands sensitive to the three-dimensional order in a polymer. These bands are relatively rare in the polyolefin spectra (see also Section III.C.2e: polyethylene and polystyrene).
2) Helix bands that disappear from the polymer melts and solutions due to the shortening of the helix sequences in the amorphous state. These bands are specific for a particular type of stereoregularity and are sensitive to the monodimensional order in a polymer (63).
3) Regularity bands having relatively low sensitivity to the aggregate state of a homopolymer but sensitive to the type of polymer stereoregularity. In the spectra of nonregular polymers and copolymers these bands become asymmetrical and their relative intensities

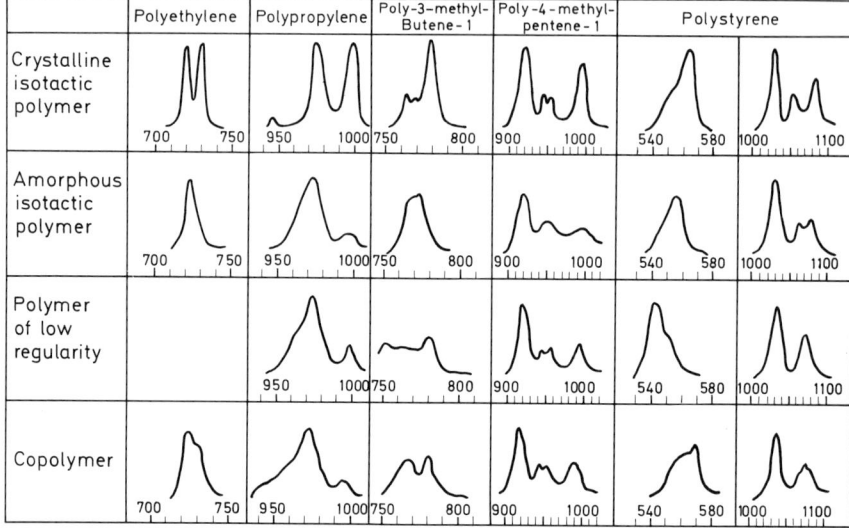

Fig. 1. IR bands for different olefins sensitive to unit distribution in copolymers (see text: Section III.C.2e)

decrease (Fig. 1). According to the theoretical approach stated above, these bands are typical examples of bands sensitive to regular block length.

c) Quantitative Methods for Evaluating Unit Distribution. The most popular quantitative method for estimating unit distribution in olefin copolymer spectra is the method of extinction coefficients with threshold sensitivity. This method was qualitatively formulated (64) and then developed by many authors (11, 53, 54, 60, 63, 65). According to these authors, one can expect a smooth dependance of K_n on n in the case of regularity bands: K_n increases with increasing n and gradually becomes constant (as well as v_n) for large n. The main assumption of the approximation used is that this smooth dependence is replaced by threshold dependence: for small n $K_n = 0$ and, beginning from any particular n, K_n becomes constant and independent of n. Possible errors in this approximation have been estimated (60).

The threshold method is widely used because the statistical dependencies for fractions of units in the sum of sterically regular blocks, beginning from any given n value (Section II), can be used for the comparison of experimental and statistical data.

Another important method of treating the spectral data is by analysis of the band shapes, which allows one to separate the complex asymmetrical band consisting of the overlapping bands that originate from blocks of different lengths on the initial bands.

d) Elimination of Conformational Defects. As mentioned above (Section III.C.2a), all three usual kinds of the polymer disorder (chemical, configurational and conformational) produce approximately the same changes in the IR spectra. Hence, if we want to estimate only the chemical and configurational disorder in a copolymer (described by the copolymer statistics given in Section II), we have to eliminate the conformational defects from the samples under investigation as completely as possible.

There are two general ways of doing this: (1) to study the IR spectra of polymer solutions at low temperatures (*54, 63, 65*); (2) to study the IR spectra of carefully crystallized products (*11, 44, 65*). Both methods yield the helical conformation for practically all isotactic segments, as evidenced from the study of polymer mixtures and block copolymers (*44, 66*). The use of the first method is severely limited for high-olefin copolymers due to their low solubility, but the second method is very convenient and has been widely applied. The simplest crystallization technique is the annealing of copolymers at suitable temperatures.

e) Sequence-sensitive Bands in the Copolymer IR Spectra. All papers dealing with the IR spectra of high-olefin copolymers include a study of the relatively small number of sequence-sensitive bands. They are briefly examined below.

Bands of Ethylene Units. All data on the distribution of ethylene units were obtained from the small spectral region 800–700 cm^{-1} (rocking modes of CH_2).

Orthorhombic polyethylene has in this region two intense, sharp bands at 720 and 731 cm^{-1} due to the splitting of the rocking modes of two chains in the crystal cell (*67*) (see Fig. 1). The spectra of linear hydrocarbons have in this region one intense, slightly asymmetric band at 722 cm^{-1}, corresponding to the isolated polymethylene chain mode. Isolated ethylene units in copolymers forming $(CH_2)_3$ blocks absorb at 733 cm^{-1} (*68, 69*). This 733 cm^{-1} band can sometimes be confused with the crystallinity band at 731 cm^{-1}, but it is wider and, unlike the crystallinity band, does not change at high temperatures (Fig. 1). $(CH_2)_2$ units absorb at 750–755 cm^{-1} (*68–70*).

Quantitative methods elaborated to calculate the absorbances of these overlapping bands include analytical (*32, 69*), computer (*71*), and compensation (*51*) techniques.

Table 1. Sequence-sensitive bands of propylene units

Position of band	Type of regularity	Classification	n	Ref.
998 cm^{-1}	isotactic	helix band	10–12	(11, 65, 72–74)
973 cm^{-1}	isotactic	regularity band	3–4	(11, 17, 56, 74, 75, 77)
841 cm^{-1}	isotactic	helix band	12	(11, 65, 72)
977 cm^{-1}	syndiotactic	helix band		(75, 76)
962 cm^{-1}	syndiotactic	regularity band		(75, 76)
867 cm^{-1}	syndiotactic	helix band		(75, 76, 78)
936 cm^{-1}	—	band of isolated propylene unit		(21, 68, 77)

Bands of Propylene Units. Most of the sequence-sensitive bands in the spectra of propylene copolymers are situated in the 1000–800 cm^{-1} region (Fig. 1). Table 1 includes the positions of these bands, their classification according to Section III.C.2b, and the lengths of blocks according to Section III.C.2c.

Bands of Butene-1 Units. The well-known phenomenon of the polymorphism of polybutene-1 has a significant influence on its IR spectrum (79–81). Some bands were found to differ for different crystalline structures, which allows identification of these structures (82). Nevertheless, the study of the IR spectra of polymer solutions showed some stereoregularity bands at 1220, 1364, 1095, 905 cm^{-1} (79, 82–84).

Bands of Pentene-1 Units. There are few papers on the polypentene-1 IR spectra (85, 86), and they deal mainly with band assignment (86) and the study of polymorphism (85). Characteristic for Modification II are bands at 1004 and 948 cm^{-1} and for Modification I that at 1047 cm^{-1}.

Bands of Styrene Units. The spectra of both isotactic and atactic polystyrene have been studied in detail (63, 65, 67, 87–92) to find sequence-sensitive bands.

These bands are shown in Fig. 1 and Table 2, together with their classifications and n values. There is a certain confusion in the assignment of the bands in the 565–558 cm^{-1} region (44, 45, 47, 94), probably due to the different behavior of these bands in the spectra of the isotactic polystyrene melt and solutions or those of styrene copolymers.

Bands of 3-Methylbutene-1 Units. Data on the IR spectra of poly-3-methylbutene-1 are very scarce (86, 99). A comparison we made between

Table 2. Some sequence-sensitive bands of styrene units

Position of the band, cm^{-1}	Type of regularity	Classification	n	Ref.
1085	isotactic	Helix or regularity band	4	(63, 65, 72)
1053	isotactic	Helix or regularity band	10	(63, 65, 72)
985	isotactic	Crystallinity band		(63, 65, 91, 92)
918	isotactic	Helix band	8–10	(63, 65, 72)
896	isotactic	Helix band	16	(63, 65, 72)
586	isotactic	Helix or crystallinity band		(92, 94, 96)
565	isotactic	Crystallinity or regularity band	4–5	(44, 45, 47, 92, 94)
558	isotactic	Helix or regularity band		(94, 96, 97)
1070	predominantly syndiotactic			(89, 92, 98)
	(in IR spectra of radical PS)			(95)
540	predominantly syndiotactic			(89, 92, 97, 98)
	(in IR spectra of radical PS)			
1075		Band of isolated styrene units		(44, 45, 47, 98)
555–550		Band of isolated styrene units		(44, 45, 47, 98)

the spectra of the crystalline isotactic sample, the polymer melt, and the atactic material revealed the presence of some stereoregularity bands. Two of these bands are specific only for the crystallized sample and may be regarded as helix bands: 778 cm^{-1} (Fig. 1) and 940 cm^{-1}. Another band sensitive to the unit distribution is the 1218 cm^{-1} band, the relative intensity of which is much lower in the spectra of amorphous samples than in those of crystalline ones.

Bands of 4-*Methylpentene*-1 *Units.* The study of the spectra of isotactic poly-4-methylpentene-1 (43, 45, 100) reveals stereoregularity bands at 1129, 997, 943, and 848 cm^{-1}. Two bands (997 and 848 cm^{-1}) are used for the study of unit distribution. According to the classification given above (Section III.C.2b), they can be regarded as helix bands. The threshold parameter n for the 997 cm^{-1} band as estimated from the IR spectra of 4-methylpentene-1 copolymers (43, 45) is 4–5.

Bands of Vinylcyclohexane Units. The IR spectrum of polyvinylcyclohexane is relatively insensitive to stereoregularity (*44*), the only regularity band being observed at 952 cm^{-1}. The study of vinylcyclohexane copolymers (*43, 44, 46*) showed that the 885 cm^{-1} band (one of the components of the strong 892–885 cm^{-1} pair assigned to the ring mode) probably corresponds to the absorption of the units in blocks, but not to that of isolated units (*43*).

D. Nuclear Magnetic Resonance

Although NMR spectroscopy is a powerful method for studying the structure of different vinyl copolymers (*101*), its usefulness for the olefin copolymers is rather limited. The main reason is undoubtedly the strong overlap between the resonances of the main-chain protons and those of the pendant groups. Unfortunately, this situation is unlikely to be improved by increasing the spectrometer frequency. One of the possible solutions to this problem is copolymerization with fully or partially deuterated monomers. The only exceptions to this unfortunate situation are ethylene–propylene copolymers and some styrene copolymers, and here important information about the copolymer structures was obtained, including data relevant to the mechanism of stereospecific polymerization (see Section IV.A.1b).

The theoretical background to NMR spectra interpretation in the case of homopolymers and copolymers has been thoroughly discussed in the excellent book by Bovey (*101*), so there is no need to dwell upon it here. We merely note that the combination of a variety of chemical and stereochemical structures in the NMR spectra of copolymers makes them so complex that "there may be relatively little advantage in the observation of such copolymers at high field strengths unless one comonomer predominates strongly, for the sequence fine structure may otherwise either overwhelm analysis or be unresolvable" [(*101*), p. 206].

Moreover, while the statistic expressions given in Section II cannot be used as such for the analysis of NMR spectra, the necessary formulas for dyad and triad populations can be easily obtained through the usual kinetic or statistical approaches [for homopolymers see (*9, 11, 101–103*), for copolymers (*24, 104*)].

The recent development of ^{13}C NMR spectrometry has not contributed significantly to the determination of high olefin structure either, due to difficulties in separating the carbon atom resonances of main chains and pendant groups. The best results have been obtained for ethylene–propylene copolymers (see Section IV.A.1b).

E. Melting-point Measurements

The theory of the melting-point depression of crystalline polymers with defects was developed by Flory (*105*) and has been applied to copolymers (*106, 107*). According to the theory, the relationship between the equilibrium melting point of a copolymer T_m and that of the corresponding homopolymer T_m^0 is:

$$1/T_m - 1/T_m^0 = -(R/\Delta H_u) \ln p,$$

where ΔH_u is the heat of fusion per crystallized unit, and p is the probability that a given stereo unit, selected at random, will be followed by the same unit. In the general case of stereospecific catalysis, the expressions for p are:

enantiomorphous model: $p = (D_1) p_{D_1 D_1} + (L_1) p_{L_1 L_1}$,
where $p_{D_1 D_1}$ and $p_{L_1 L_1}$ are defined by Eq. (5) and (D_1) and (L_1) by Eq. (4);
racemic model: $p = p_{D_1 D_1} = p_{L_1 L_1}$ (see Eq. (5) for $p_{D_1 D_1}$).

In the case of ideally stereospecific catalysts and in the case of symmetrically substituted olefins (ethylene),

$$p = p_{M_1 M_1} = r_1 F/(1 + r_1 F) \quad \text{[see Eq. (9)]}.$$

When these expressions for p are combined with Eq. (12), it is evident that for the given kind of units the melting-point depression depends on the copolymer composition (f), on the stereospecificity of the catalyst (R'), and on the product of the reactivity ratio ($r_1 r_2$).

If we take an ideally specific system ($R \to \infty$) and random copolymerization ($r_1 r_2 = 1$), all expressions for p are reduced to

$$p = f/(1 + f) = C_{M_1}$$

i.e. a mole fraction of a given monomer in a copolymer.

In such a case the depression of the equilibrium melting point of a copolymer is very sensitive to its composition. For example, for such propylene copolymers [$T_m^0 = 170°C$, $\Delta H_u = 2370$ cal/mole (108)] the melting points are (109):

C_{M_1}	100%	90%	80%	70%
T_m	170°C	155°C	135°C	120°C

If there is a tendency to block formation ($r_1 r_2 > 1$), the melting points of copolymers will be higher, and if there is a tendency to alternating addition ($r_1 r_2 < 1$), they will be lower than for random copolymers.

Experimental tests of the theory have shown that in some cases the correlation between theory and experiment is good (110), but sometimes deviations from dependence occur, and these can be both positive (110) and negative (107, 111); the latter may be attributed to the non equilibrium conditions of the crystallization of long blocks (107) and to the difference between the ΔH values for copolymers and homopolymers (111). At any rate, the data on melting point depression make it easy to differentiate between random and block copolymers of the same composition (111).

In practice, there are two other kinds of limitation in the application of this method to copolymers:

1) When the second comonomer incorporates the crystallites of the first comonomer, melting-point behavior changes significantly and can sometimes lead to an increase of the melting point with the growth of the content of the second monomer [see Section III.F. and (112)].
2) The procedure for measuring of the equilibrium melting point is very complicated (107, 110) and is sometimes replaced by more rapid procedures (DTA analysis) which reduce the validity of comparisons between experimental and theoretical data.

F. Crystallinity Measurements

Quantitative estimation of copolymer crystallinity by the X-ray method, dilatometry, etc., is a useful tool for characterizing unit distribution in copolymers.

Generally speaking, the "dilution" of any crystalline polymer by certain defects will sharply decrease its crystallinity, due to the shortening of the regular sequences. A good example is ethylene–vinyl acetate copolymers. The crystallinity of polyethylene is about 80–90%, but the incorporation of 2 mol-% of vinyl acetate reduces it to ~60%, and over 10% of vinyl acetate completely destroys the crystallinity (113). Another estimate of this "dilution" effect gives the ratio 1 : 2 (i.e. ~30% of "foreign" units) as the limit of the region of the completely amorphous copolymer (114). From this point of view, if both olefins produce crystalline homopolymers, we can expect a U-shaped curve in the plot of "total crystallinity versus copolymer composition".

This simple situation is frequently altered by (1) isomorphism phenomena and (2) the inhomogeneity of copolymer composition (see Section V).

The theoretical considerations and the experimental examples of isomorphism in the case of olefin copolymers have been carefully discussed (112, 114). This phenomenon is not directly related to the problems discussed in this review [see (112, 114, 115) for definitions and examples]. It is worth mentioning only that the majority of isotactic polyolefins have similar helices (3_1-4_1) and pendant groups of similar volume, which significantly favors the isodimorphism of their copolymers (114). Disregard of this phenomenon can lead to a serious overestimation of the block content of olefin copolymers, so that a careful evaluation of possible isomorphism is necessary in every case.

The statistical parameters that qualitatively describe copolymer crystallinity are the portions of units in the sum of long, stereoregular blocks [Eqs. (8) and (15)].

The main practical method for measuring crystallinity is based on X-ray diffractometer scans of copolymers and is similar to that used for polypropylene (93). It consists of the division of the area under the peaks by linear baselines into amorphous (A) and crystalline (C) areas, the degree of crystallinity being $C/(C + A)$.

IV. Structure of High-olefin Copolymers

This section contains experimental results relating to the structure of different copolymers of high olefins. All data are listed under the main olefins used for copolymerization: ethylene, propylene, butene-1, 3-methylbutene-1, 4-methylpentene-1, and styrene. When such a classification is used, it is inevitable that some copolymers will be mentioned twice (e.g. ethylene–styrene and styrene–ethylene copolymers). Where this occurs, the data on copolymer structures are given under the first heading and the second heading refers to the first. For example, all data about ethylene–styrene copolymers are cited under "Ethylene copolymers", subheading "Ethylene–styrene copolymers", and under "Styrene copolymers" there is only a reference to the earlier subheading.

A. Ethylene Copolymers

1. Ethylene–Propylene Copolymers

Ethylene–propylene copolymers have important commercial applications, so that an enormous number of papers have been and are still being published on all aspects of the synthesis, characterization and properties of these products. Some of these results have been reviewed [see Chapter III to VI in (2)].

Recently an excellent review (32) appeared containing exhaustive information about IR studies of ethylene–propylene copolymers, including methods of determining composition by different spectral methods, and the problem of monomer distribution in the copolymers. For this reason, we do not discuss here methods of determining copolymer composition and we mention other aspects only briefly.

a) Chemical Structure of the Copolymers. An examination of the data on ethylene–propylene copolymers leads to the conclusion that from the chemical point of view the copolymerization process is very regular. Both comonomers polymerize in 1,2-position and mostly head-to-tail addition of propylene units takes place; all other possible modes of polymerization (1,3-, 2,2-, and 1,1-polymerization of propylene, etc.) can be excluded in the great majority of cases. In copolymers with a low ethylene content practically all the ethylene units are isolated and form —$CH(CH_3)$—CH_2—CH_2—CH_2—$CH(CH_3)$ sequences, which are easily detected by the presence of the 733 cm^{-1} band in the copolymer spectra (see Section III.C.2e) (32). On the other hand, when the propylene content is very low, all its units are isolated and are characterized by the 935 cm^{-1} band (32). When the catalytic system $(C_5H_5)_2TiCl_2$—$Al(C_2H_5)_2Cl$ (which cannot polymerize propylene) is used for copolymerization, all the propylene units are isolated, regardless of the copolymer composition (116).

The only important chemical deviation is the presence of $(CH_2)_2$ units in the copolymers (117, 119–123). There are two ways in which these blocks can form in the copolymers (120, 123):
1) head-to-head propylene addition [in this case the —$CH(CH_3)$—
   —$CH(CH_3)$-units present in chains are manifested in the ~1130 cm^{-1}

band (*123*)]. The portion of monomer inversions in polypropylene is of the order of 4–7% in the case of vanadium systems (*120*) and is very low for titanium systems;

2) formation of isolated ethylene units surrounded by two propylene units, one of which is in normal and the other in reversed position (*120, 123*); —CH$_2$—CH(CH$_3$)—CH$_2$—CH$_2$—CH(CH$_3$)—CH$_2$—.

The relative probabilities of these two processes were estimated (*120*) and from them some sophisticated conclusions were drawn concerning the mechanism of the insertion reactions with the syndiospecific catalyst VCl$_4$—Al(C$_2$H$_5$)$_2$Cl (*123*).

The presence of (CH$_2$)$_2$ units in ethylene–propylene copolymers (as well as in polypropylene) is characteristic mainly for the products obtained with vanadium-containing systems, whereas the products obtained with titanium-containing catalysts are practically free from them (*120, 122, 124*).

b) Unit Distribution in Ethylene–propylene Copolymers. The distribution of monomer units in these copolymers has been examined by a variety of physical methods (IR, NMR, X-rays, etc.). The $r_1 r_2$ values have been measured for many systems [see reviews and Refs.(*2, 127–129*)].

In the case of copolymers obtained with isospecific catalytic systems, both qualitative (*125*) and quantitative IR data (*71, 118*) demonstrate that the distribution of ethylene units roughly corresponds to that expected on the basis of the $r_1 r_2$ values[1]. Some of the possible deviations are due to the presence of isolated ethylene units as (CH$_2$)$_2$ sequences.

The IR estimations of the distribution of the propylene units cover two regions of copolymer composition: the region of high propylene content, where the most convenient value to be measured is the fraction of propylene units in the isotactic blocks of different lengths (*10, 72, 122*), and the region of relatively low propylene content, where the convenient parameter is the ratio between the isolated units and the units in blocks (*21, 121, 126, 131*) (for band assignments, see Table 1).

The general conclusion is that the distribution of the propylene units also agrees with the copolymer statistics. The distribution of propylene units in the sums of blocks of different lengths in the copolymers obtained with the TiCl$_3$—Al(C$_2$H$_5$)$_3$ system is shown in Fig. 2. Confidence in the correlation between the real monomer distribution in copolymers and

---

[1] A significant disagreement between the statistics and IR data was found (*122*) for some copolymers obtained with TiCl$_3$-containing systems. This disagreement is probably partly due to the misassignment of the sharp crystalline band at 731 cm^{-1} to the absorption of isolated ethylene units (see Section III.C.2e). The presence of long ethylene sequences in these copolymers was confirmed by NMR in the same paper.

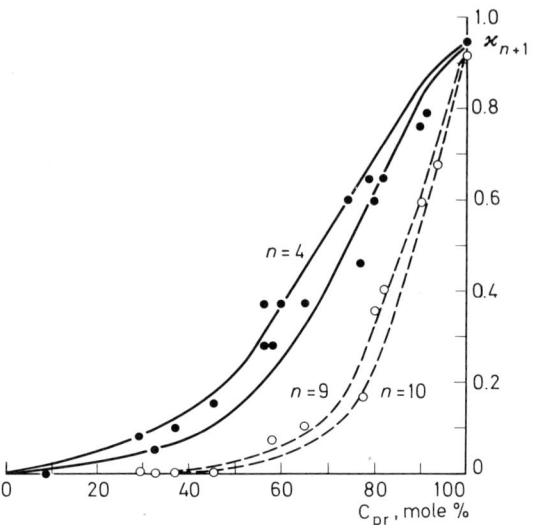

Fig. 2. Distribution of propylene units in ethylene–propylene copolymers in the isotactic sequences of different lengths (72). Catalytic system $TiCl_3$—$Al(C_2H_5)_3$. ● = data for the 973 cm^{-1} band, ○ = data for the 998 cm^{-1} band. The calculations are carried out with Eqs. (8) and (12) for $n = 4$ [$r_1 r_2 = 1.7$ and 0.8 (53)] and for $n = 9$ and 10 ($r_1 r_2 = 1$)

the $r_1 r_2$ values is so strong that some authors use it to obtain $r_1 r_2$ values from IR data alone (21).

One of the features of syndiospecific catalytic systems is a strong tendency for monomer units to alternate in copolymerization (126) in accordance with the low $r_1 r_2$ values (0.26–0.28). This tendency is confirmed by the large fractions of isolated monomer units in copolymers. Ethylene units exist in such cases as $(CH_2)_3$ and $(CH_2)_2$ sequences (123) and dominate even in copolymers with 1 : 1 composition (122). The ratio between the isolated propylene units and those in blocks as measured by the IR method (126) is also significantly higher in this case than for copolymers obtained with nonspecific catalytic systems ($r_1 r_2 \sim 1$), thus confirming the tendency to alternation.

A 220-MHz PMR study of ethylene–propylene copolymers is reported (122) and (132). A comparison between the spectra of polypropylene (133) and that of hydrogenated rubber (the model for the alternating copolymers) (122) enabled the authors to assign some resonances in the copolymer spectra, but the assignment of some peaks remains dubious and precludes direct comparison with statistics.

The difference between the shapes of CH_3 resonances of two copolymers of practically the same composition, one obtained with the highly isospecific system $TiCl_3$—$Al(C_2H_5)_2Cl$ and the other with the nonspecific system $VOCl_3$—$Al(C_2H_5)_2Cl$, qualitatively confirms the quantitative conclusion deduced from the IR data (*10*) that the configuration of propylene sequences in the first copolymer exhibits much more regularity than that in the second (*132*).

Important conclusions about the structure of ethylene–propylene copolymers and the mechanism of stereocontrol were obtained from C^{13} NMR studies of these copolymers (*133–135*). A detailed assignment of all resonances in these complex, strongly overlapping spectra was made (*135*), allowing the estimation of the $r_1 r_2$ value for copolymers obtained with the V acetylacetonate—$Al(C_2H_5)_2Cl$ system.

References (*133*) and (*134*) deal mainly with the C^{13} NMR spectra of copolymers with low (10–20%) ethylene content. Their main conclusions concern the mechanism of stereocontrol:
1) in the case of isospecific catalytic systems, stereoregulation is due to the dissymmetry around the metal atom in the active site and is not influenced by the structure of the last unit of the growing chain, *i.e.* it is enantiomorphous [only $(CH_2)_3$ units exist in these copolymers];
2) in the case of syndiospecific catalytic systems, stereoregulation is exerted predominantly by the last unit in the growing chain (racemic mechanism).

Certain studies are concerned with crystallinity and melting-point measurements in ethylene–propylene copolymers (*129, 136–141*) and in model copolymers (*107, 110*). Here the main conclusion is that the insertion of even a small number of propylene units in polyethylene chains leads to a sharp reduction of the polyethylene crystallinity from 70–80% for pure polymer down to 25–30% for $\sim 25\% C_{pr}$ copolymers, and to practically amorphous products for $\sim 35\% C_{pr}$ copolymers (*136, 137*). It was found (*141*) that, for copolymers characterized by different $r_1 r_2$ values, the dependence of polyethylene crystallinity on copolymer composition correlates well with the calculated dependence of the probability of the occurrence of long ethylene sequences (the properly normalized value of $\sum_{n}^{\infty} \delta_n/n$) on copolymer composition. It is thus possible to estimate roughly that the minimal effective size of the ethylene sequence able to crystallize (n_{cr}) is 8. Similar results were obtained ($n_{cr} \sim 10$) for ethylene–propylene–butene-1 terpolymers with which the fuming nitric acid digestion method was used for crystallinity measurements (*138*).

It is expedient to regard these results as qualitative rather than quantitative, bearing in mind the complex mechanism of copolymer crystallization (107, 110) and the difficulties involved in a direct comparison between the unit distribution in isolated copolymer chains and the three-dimensional packing of defective chains.

An original chemical method elaborated for the quantitative study of the monomer unit distribution (137, 142) consists of copolymer pyrolysis with subsequent gas-chromatograph analysis of the decomposition products. Good correlation was found between the experimentally measured frequency distribution of fragments of copolymers with up to 50 mol-% of propylene and those calculated from $r_1 r_2$ values [system $VOCl_3$—$Al_2(C_2H_5)_3Cl_3$, $r_1 r_2 = 0.87$] (142).

2. Ethylene–Butene-1 Copolymers

These copolymers have mechanical properties similar to those of ethylene–propylene copolymers and were also the subject of detailed studies [see (2), Chapter VII].

a) Determination of Copolymer Composition. Two main experimental methods were applied, based on C^{14}-labeled monomers (36, 126), and IR methods (34, 49, 51). Table 3 outlines the IR methods.

b) Chemical Structure and Unit Distribution. The presence in the ethylene–butene-1 IR spectra of bands at 722 and 733 cm^{-1} (51), assigned to the $(CH_2)_{n \geq 5}$ and $(CH_2)_3$ rocking modes respectively, reveals the predominance of the usual head-to-tail addition as in ethylene–propylene copolymers; and the presence of the 770 cm^{-1} band, corresponding to ethyl group vibration (51, 143), shows that butene-1 units incorporate copolymer chains without isomerization. The same is true for the ethylene–butene-1 copolymerization with the soluble $(C_5H_5)_2TiCl_2$—$Al(C_2H_5)_2Cl$ system in which no homopolymerization of butene-1 takes place, and all butene-1 units in the copolymers are isolated (144).

The crystallinity of the copolymers has been measured by different methods: X-rays (49, 138–140, 145–147), density measurements (140), and IR methods (with 731 and 1894 cm^{-1} bands) (49, 144). Some qualitative conclusions may be deduced about the unit distribution. These data are combined in Fig. 3 and demonstrate a significant diminution of polyethylene crystallinity as the butene-1 content increases, thus proving that random copolymerization takes place[2]. In the case of a

---

[2] The crystallinity values (138) obtained with the fuming nitric acid digestion method for the copolymers synthesized with a Ti-containing system ($r_1 r_2 = 1.5$) are slightly higher: 75% for 3.4% of butene-1 and 44% for 13.7% of butene-1 in the copolymers.

Table 3. IR methods for determining the composition of ethylene–butene-1 copolymers

Analytical bands	Calibration method	Analytical expression for the calibration curve; samples	Measurement range and precision	Ref.
5910 cm^{-1} – 5680 cm^{-1} – the first overtones of CH$_3$ and CH$_2$ stretching modes	Standards with C^{14}-ethylene	$A_{5910}/A_{5680} = 0.018 C_{but} + 0.01$ films, ~1 mm	20–90% ±1%	(34)
2959 cm^{-1} – 2854 cm^{-1} – CH$_3$ and CH$_2$ stretching modes	Standards with C^{14}-ethylene	$A_{2959}/A_{2854} = 0.023 C_{but} + 0.02$ 0.3% solns in CCl$_4$, cell ~1 mm	20–90% ±1%	(34)
1380 cm^{-1} – CH$_3$ sym. deform. mode	Absorption coefficient in IR spectra of hydrocarbons	$N_{C_2H_5}/100 C = 410 A_{1380}/l(\mu)$ films, ~70 μ compensation PE	0–20%	(49)
770 cm^{-1} – ethyl group mode	Absorption coefficient in IR spectra of hydrocarbons	Linear correlation between $N_{C_2H_5}/100 C$ and A_{770}/l	0–20%	(49)
772 cm^{-1} 722 cm^{-1} 733 cm^{-1} (see III.C.2e)	Absorption coefficient in IR spectra of hydrocarbons	$C_{but}(\text{mol-}\%) = \dfrac{1.3 A_{772}/A_{722}}{1 + 0.813 A_{733}/A_{722} + 0.65 A_{772}/A_{722}}$ films at 120° C, compensation		(51)
1375 cm^{-1} 770 cm^{-1}		Branching determination films, PE compensation		(143)
1380 cm^{-1} 1460 cm^{-1}	Standards with C^{14} monomers	Analytical curve A_{1380}/A_{1460} versus (CH$_3$)/((CH$_3$) + (CH$_2$)) ratio, films, 120° C		(138)

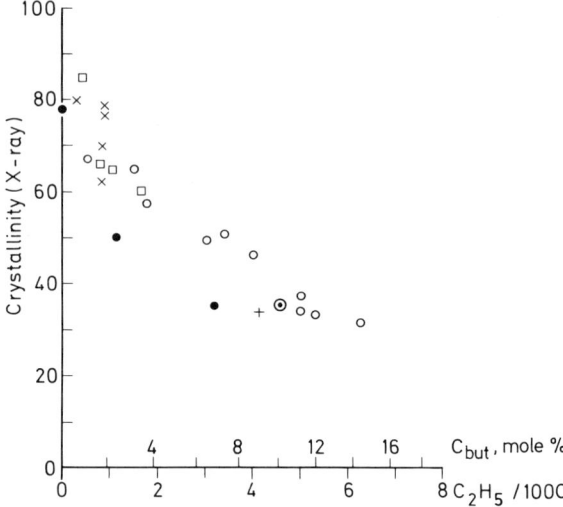

Fig. 3. Crystallinity of ethylene blocks in ethylene–butene-1 copolymers. The data are from: ○ (49), ● (139), □ (140), + (145), × (146), ⊙ (147)

V-containing catalytic system, products having a butene-1 content of more than 30% are amorphous (147).

The decline in crystallinity is accompanied by an increase in the polyethylene lattice spacing (139, 140), due to the influence of the side groups on the packing of chains. Ethyl branches have more influence on crystallinity than do methyl branches (49, 139, 140).

The data on the melting points of the copolymers presented in Ref. (140) cover only a small composition range. The melting points are lowered from 134.5 to 128.6° C as the butene-1 content increases from 0 to 3.5%, and the width of the DTA peaks increases significantly in this range, indicative of real copolymer formation. Reference (145) deals with the ethylene–butene-1 copolymers obtained with the highly isospecific system $TiCl_3$—$Al(C_2H_5)_2Cl$. From a combination of crystallinity and melting-point measurements, the author asserts that significant phase separation takes place, yielding a mixture of ethylene and butene-1 enriched, partly crystalline products together with the amorphous random copolymer (see Section V).

An interesting example of ethylene–butene-1 copolymer formation (146) occurred when ethylene was polymerized with the catalytic systems $TiCl_4$—$Al(C_2H_5)Cl_2$ and $TiCl_3$—$Ti(OC_2H_5)_4$—$Al(C_2H_5)_3$, which evolve butene-1 due to the ethylene dimerization. The formation of ethylene–butene-1 copolymers containing up to 3 mol-% of butene-1 was proved by different methods.

Finally, we mention the crystallinity measurements in ethylene–propylene–butene-1 terpolymers (138). The catalytic system used provides relatively poor randomness ($r_1 r_2 = 2.5$) but the incorporation of a small amount of butene-1 in copolymers (5–10 mol-%) significantly reduces the total crystallinity.

3. Ethylene–Butene-2 Copolymers

It is well known that the usual Ziegler-Natta catalysts do not directly polymerize β-olefins but that they can copolymerize them with some α-olefins. A good example of this phenomenon is the ethylene–butene-2 copolymerization, which has been thoroughly studied (13). The butene-2 content of the copolymers was measured radiochemically or by the IR method with the 1380 cm^{-1} band. The absence of significant isomerization of butene-2 to butene-1 was proved by the absence from the copolymer spectra of the 773 cm^{-1} band (ethyl branching).

In this copolymerization $r_2 = 0$, and it is characterized by the limit content of butene-2 in the copolymers [see Eq. (2)]: when $F \to 0 f \to 1$ ($C_{but-2} \to 50\%$). According to statistics, copolymers with $r_1 r_2 = 0$ and $f \to 1$ are alternating, i.e. $\delta_1(A) \to 100\%$, and $\delta_1(B) \to 100\%$ [see Eqs. (10) and (12)]. These alternating ethylene–butene-2 copolymers were obtained in practice after the fractionation of copolymers obtained at low F ratios, and they contained 47–48 mol-% of butene-2. The IR study confirmed the alternating structure of these products. The main band in their spectra in the region of CH$_2$ rocking modes is that at 753 cm^{-1} corresponding to (CH$_2$)$_2$ sequences, i.e. isolated ethylene units. These copolymers are excellent models of the head-to-head polypropylene structure.

4. Ethylene–Higher Linear Olefin Copolymers

The only published IR method for measuring the ethylene–pentene-1 copolymer composition (148) uses the absorbance ratio of the 722 and 739 cm^{-1} bands, the former band being assigned to long methylene sequences and the latter to propyl branching. Compensation with n-heptadecene is recommended to separate the two overlapping bands, and the analytical expression is $C_{pent}(\%) = 35 A_{739}/A_{722} + 1.6$ (region of measurement 10–30% of pentene-1, precision 3.3%). The presence of the 739 cm^{-1} band (typical for propyl groups) proves that no isomerization of pentene-1 occurs during copolymerization with ethylene.

Some qualitative data prove these copolymers to be random. It is necessary to incorporate ~20 mol-% of pentene-1 into the copolymer to obtain amorphous products (*149*); copolymers with a lower pentene-1 content exhibit low crystallinity (*148, 150*). The melting points of model copolymers obtained by diazoalkane synthesis (*110*) demonstrate a sharp fall with increasing branch content from 113.7° C for 2.0% to 89.0° C for 4.2%, and 64° C for 6.4%.

Ethylene–hexene-1 copolymers were also studied (*163*). The composition measurements were made by the IR method. The 1380 cm^{-1} methyl deformation band was used as a measure of the hexene-1 content and the 4310 cm^{-1} band as the internal thickness standard, the analytical expressions being C_{hex}(wt.-%) = 15.225 A_{1378}/A_{4210} − 1.26 and C_4H_9/100C = 2A_{1378}/A_{4310} + 0.5.

The copolymers obtained are random ($r_1 r_2 = 0.58$). The polyethylene crystallinity sharply decreases as the hexene-1 content grows being ~25% at 10 mol.-%, ~15% at 15 mol.-%, and ~5% at 20 mol.-% of hexene-1; copolymers containing more than 25% of hexene-1 are amorphous.

5. Ethylene–3-Methylbutene-1 Copolymers

Very few data are available on the copolymerization of this monomer pair (*149, 152*). The copolymers were obtained with the V acetylacetonate—Al(isoC$_4$H$_9$)$_2$Cl catalytic system at 10–30° C. The introduction of 3-methylbutene units increases the solubility of the copolymerization products and decreases their melting points, which is an indication of random copolymerization ($r_1 r_2 = 0.24$):

3-methylbutene-1 content, %	7.5–7.9	14.9	23.4
Solubility in *n*-heptane, %	61–65	77.4	81.5
Melting point, °C	127–129	110	90

6. Ethylene–4-Methylpentene-1 Copolymers

The general IR method for determining the composition was elaborated in (*34*) and is based on the intensities the bands at 920 cm^{-1} (isopropyl group mode) and 4310 cm^{-1} (internal thickness standard). The analytical expression is $A_{920}/A_{4310} = 0.038 C_{4mp1} + 0.035$ and it is valid in the 0–20 C_{4mp1}% region (precision ±1%).

We studied the IR spectra of the copolymers obtained with the highly isospecific catalytic system $TiCl_3$—$Al(C_2H_5)_2Cl$. In this case the chemical structure of the copolymers is predominantly normal, i.e. 1.2-head-to-tail addition of olefin units, as is proved by the absence of $(CH_2)_{n=2}$ bands from the spectra of poly-4-methylpentene-1 (only a very weak band is present at 745 cm^{-1}, probably due to head-to-head addition). There is only one weak band at 730–735 cm^{-1} in the spectra of copolymers of low ethylene content, due to the $(CH_2)_3$ sequences.

The opposite is true of ethylene–4-methylpentene-1 copolymers obtained with the catalytic system V acetylacetonate—$Al(isoC_4H_9)_2Cl$ (153). These copolymers contain both units from normal 1,2-addition and isomerized units of the type —CH_2—CH_2—CH_2—$C(CH_3)_2$—, characterized by the 1230 cm^{-1} band.

The distribution of 4-methylpentene-1 units in copolymers obtained with $TiCl_3$—$Al(C_2H_5)_2Cl$ was studied by the IR method, using the 997 cm^{-1} band absorbance as the measure of this olefin fraction in long

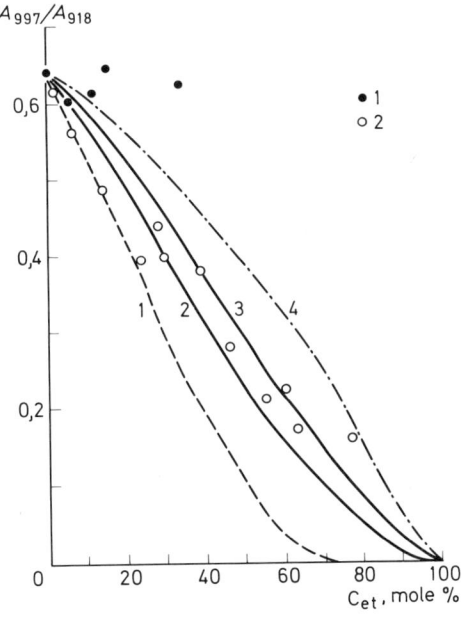

Fig. 4. Distribution of 4-methylpentene-1 units in ethylene–4-methylpentene-1 copolymers. 1. products of the successive polymerization of 4-methylpentene-1 and ethylene; 2. copolymers. Calculations are carried out with Eqs. (8) and (12) for $r_1 r_2 = 1$, $n = 4$ (4); $r_1 r_2 = 5$, $n = 5$ (2); $r_1 r_2 = 5$, $n = 4$ (3); and $r_1 r_2 = 10$, $n = 4$ (4)

isotactic sequences (see Section III.C.2e) normalized to the 918 cm^{-1} absorption (Fig. 4). The statistical curves in Fig. 4 are the products $(A_{997}/A_{918})_{P4mp1} \cdot \varkappa_{n+1}$. The value $(A_{997}/A_{918})_{P4mp1}$ was measured in the spectra of highly isotactic poly-4-methylpentene-1 and $\varkappa_{n+1}$ was calculated for $n = 4$ and $n = 5$ and for different $r_1 r_2$ values using Eqs. (8) and (12). It is evident from Fig. 4 that the distribution of 4-methylpentene-1 units in the copolymers corresponds to the $r_1 r_2$ value ~ 5, i.e. this copolymer has a strong tendency toward block formation. This conclusion is qualitatively supported by the examination of the 730–720 cm^{-1} region in the IR spectra. All copolymers with ethylene content over 10% have the 731–720 cm^{-1} doublet, indicating the presence of long crystalline ethylene blocks, and the band of isolated ethylene units at 733 cm^{-1} appears only in the spectra of copolymers with 7–5 mol.-% of ethylene. This tendency toward block formation finds support in the $r_1 r_2$ value measured from the copolymer composition data (~ 4.8).

In the case of copolymers obtained with the V acetylacetonate––Al(isoC$_4$H$_9$)$_2$Cl system, the $r_1 r_2$ value is much smaller [~ 0.7 from (153)], and $\sim 20\%$ of 4-methylpentene-1 in the copolymer is sufficient to make the product amorphous.

7. Ethylene–Styrene Copolymers

The main structural results on these copolymers were obtained for products of low styrene content (41, 154) polymerized with Ti-based systems. In all cases the total polymer products can be separated into two parts: atactic polystyrene (soluble in ketones) and the true copolymers. These products do not contain isotactic polystyrene (41), which verifies the suggestion that this atactic polystyrene is formed on cationic active sites rather than on the usual Ziegler-type centers. Copolymers of low styrene content have the styrene units isolated independently from the $r_1 r_2$ value. This was proved by IR and NMR spectra studies. Styrene units absorb at frequencies characteristic for isolated groups at 550–560 cm^{-1} and at 1075 cm^{-1} when the styrene content is 5.7–19.1% (41, 154), and give rise to a resonance at τ 2.98 in the NMR spectrum (154).

The ethylene units in such copolymers are obviously organized in long crystallizable sequences producing a narrow crystallinity band at ~ 730 cm^{-1}. The IR method of copolymer analysis could therefore be based on the A_{760}/A_{730} ratio and is suitable for the 0–20 mol.-% range of styrene content (41).

B. Propylene Copolymers

1. Propylene–ethylene Copolymers (see Section IV.A.1)
2. Propylene–Deuteropropylene Copolymers

IR spectra were studied of a 1:1 propylene–propylene-1,1,2-D_3 copolymer obtained with the $TiCl_3$—$Al(C_2H_5)_3$ catalyst (65, 155). This copolymer is highly crystalline (from X-ray diffraction) due to the isomorphous substitution of one monomer unit for another in the polymer helix. The IR spectrum of the copolymer is similar to the superposition of the spectra of molten polypropylene and that of polypropylene-1,1,2-D_3. The bands at 1166, 998, and 841 cm^{-1} are absent, and the 973 cm^{-1} band has evident skewing on the long-wavelength side. These results support the validity of the application of these bands to the evaluation of propylene unit distribution and allow estimation of the minimum sequence length necessary for the 998 cm^{-1} band to appear in the spectra of the propylene copolymer (see Table 1).

3. Propylene–Butene-1 Copolymers

a) Determination of the Copolymer Composition. There are two main experimental methods for the measurement of the composition of these copolymers, *i.e.* radiochemical and IR. The IR methods are listed in Table 4.

b) Copolymer Structure. The monomer unit distribution was found to depend significantly on the catalytic system used. With highly isospecific systems, such as $TiCl_3$—$Al(C_2H_5)Cl_2$-hexamethylphosphoric triamide [$r_1 r_2 = 3.44$ in the presence of H_2 (156)] or $TiCl_3$—$Al(C_2H_5)_2Cl$ (145), highly blocky copolymers were formed, having two distinct melting points (for copolymers containing 3 to about 80% of butene-1) and X-ray crystallinity characteristic of two homopolymers. Some of these copolymers were obtained with very low conversions (1–6%) (156) and have a relatively narrow range of composition; the blockiness was thus due rather to the specific unit distribution than to the wide compositional distribution. The melting-point values (156, 145) confirmed that real copolymers were formed:

propylene content, %	100	90–89	85–75	44
melting point, °C	175	152–147	142–138	100

Table 4. IR methods for determining the composition of propylene–butene-1 copolymers

Analytical bands	Calibration method	Analytical expression for the calibration curve, samples	Measurement range and precision	Ref.
766 cm^{-1} – ethyl group band	Standards with C^{14}-butene-1. For polybutene-1 $A_{766}/l(\text{cm}) = 77$	$C_{but}(\text{wt.-}\%) = 510 A_{766}/l$ (mill) pressed films, 100 μ	0–100 % ±2%	(33)
1145 cm^{-1} – band of both monomer units 772 cm^{-1}	Radiochemical standards	A_{1145}/A_{772} $= 0.705 C_{pr}/C_{but} + 0.244$ molten films at 160° C, ~0.1 mm	5–25% of butene ±2%	(34)
809 cm^{-1} or 1150 cm^{-1}, 764 cm^{-1}	Homopolymer blends, standards with C^{14}-propylene			(37) (156)
765 cm^{-1}, 974 cm^{-1} – propylene unit band	Homopolymer blends	Calibration curve is given for A_{765}/A_{974} versus mol.-% of butene-1	0–90% of butene	(38)
767 cm^{-1}, 4310 cm^{-1} – internal thickness standard	A_{767}/A_{4310} in the polybutene spectra is 4.2	$C_{but}(\text{wt.-}\%) = 23.8 A_{767}/A_{4310}$ films, 0.1 mm		(126)

Unfortunately, no comparison was made with the Flory equation (Section III.E). The changes in the cell dimensions can also be due to the inclusion of "foreign" units in the copolymer chains (145).

With less stereospecific catalytic systems, like $TiCl_3$—$Al(C_2H_5)_3$ and VCl_3—$Al(C_2H_5)_3$ (38, 157), probably more random copolymers were formed, judging by the qualitative data presented. Thus, when going from polypropylene to the copolymer with $C_{pr} \sim 80\%$ (157) [the copolymer composition from Fig. 1 in (38)], the relative absorbance of the 998 cm^{-1} helix band becomes only half as much, and qualitatively resembles the ethylene–propylene copolymers [see Fig. 7 and (72)].

These copolymers are crystalline but the d-spacing differs somewhat from those of both homopolymers (157).

A new band was found in the IR spectra of these copolymers at 1240 cm^{-1} (38) with the maximum absolute intensity in the 1:1 copolymer spectra and was considered to be characteristic for propylene–butene-1 random copolymers.

When propylene–butene-1 copolymers were obtained with syndiospecific catalytic systems at low temperatures ($r_1 r_2 \sim 0.5$), the tendency to alternation of monomer units was significant (126). These copolymers

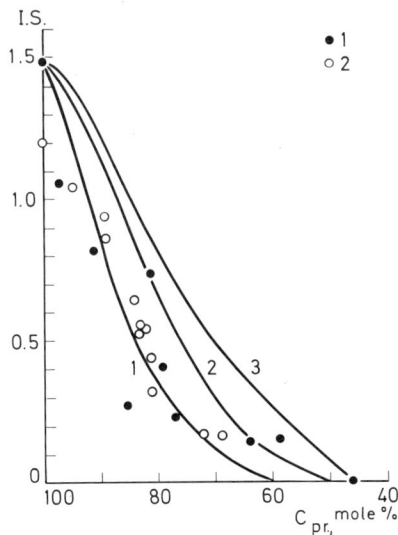

Fig. 5. Distribution of propylene units in propylene–butene-1 copolymers in syndiotactic sequences. Experimental data for the syndiotacticity index (I.S.) are from (126) for catalytic systems VCl_4—anisol—$Al(C_2H_5)_2Cl$ (1) and VCl_4—$Al(C_2H_5)_2Cl$ (2). The calculations are carried out with Eqs. (15) and (12) for $n = 10$ (1), $n = 7$ (2) and $n = 5$ (3) for $r_1 r_2 = 0.5$

are partly crystalline according to the X-ray data; this is thought to be due to the isomorphism phenomenon: "in syndiotactic chains CH_3 groups can be replaced by C_2H_5 groups to a fairly high extend without completely preventing crystallization" (126). The IR spectroscopic parameter "syndiotacticity index" (I.S.) of propylene units calculated in the copolymer spectra is plotted in Fig. 5 versus copolymer composition [from (126)]. This parameter was determined with the formula (158) I.S. $= A_{867}/0.5(A_{4310} + A_{4260})$, the 867 cm^{-1} band being the syndiotactic helix band (Table 1), and the 4310 and 4260 cm^{-1} bands being internal thickness standards. These latter bands are also present in the polybutene-1 IR spectrum (126); so the I.S. parameter calculated from the copolymer spectra can be roughly defined statistically as (I.S.)$_{cop}$ = (I.S.)$_{max} \cdot \varkappa_{(syndio)n+1} \cdot C_{pr}$, where (I.S.)$_{max}$ corresponds to the maximum value of the parameter [1.85 from (158)] and $\varkappa_{(syndio)n+1}$ is determined by Eq. (15). The necessary calculations were made by us for different n values, for $r_1 r_2 = 0.5$, and for the relative syndiotacticity of the catalytic system $= 1.50/1.85 = 0.81$ [I.S. $= 1.5$ for polypropylene obtained with the same catalytic system as the copolymers in (126)]. These calculated data are also presented in Fig. 5. It is evident from the comparison that the experimental data correlate well with the statistics, demonstrating the random structure of the propylene–butene-1 copolymers obtained with syndiospecific systems. The minimum sequence length sufficient for the 867 cm^{-1} band to appear in the IR spectrum of syndiotactic polypropylene is of the order of 10 units.

4. Propylene–Higher Linear Olefin Copolymers

Some data have been published on the copolymers of propylene with different linear high olefins (33, 40). The main method of analysis is based on the 720 cm^{-1} band absorbance in the IR spectra (calibration with homopolymer blends). For propylene–butene-1-octene-1 copolymers the dependance is (33):

$$C_{oct}(\text{wt.-\%}) = 539 A_{719}/l(\text{mill}) \text{ in the } 0{-}25\% \text{ range}.$$

The introduction of linear olefin units (decene-1 and pentadecene-1) into the polypropylene chain significantly reduces the product density (which means a decrease in polymer crystallinity) and strongly influences some mechanical properties of polypropylene (40).

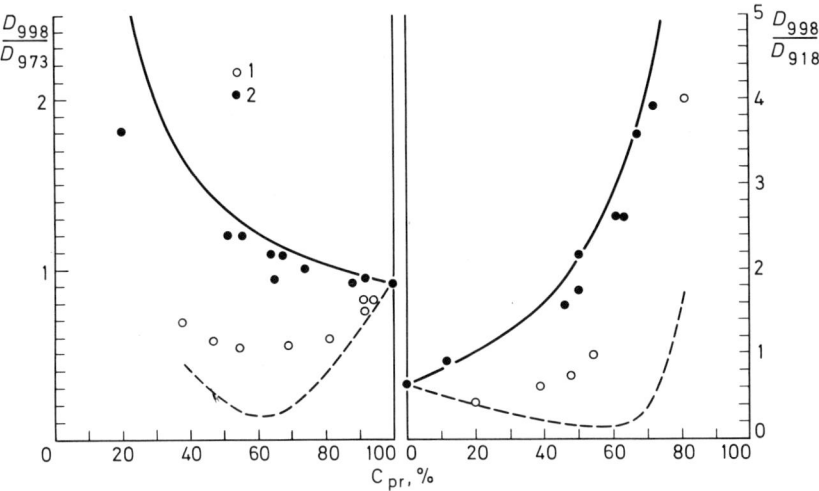

Fig. 6. Distribution of monomer units in propylene–4-methylpentene-1 copolymers (66). Experimental points (see text) are for the products of the successive polymerization (1) and for copolymers (2). Calculations are carried out for homopolymer mixtures (upper curve) and for random copolymers (lower curve)

5. Propylene–4-Methylpentene-1 Copolymers

These copolymers obtained with the $TiCl_3$—$Al(C_2H_5)_2Cl$ system were studied by IR, X-ray, and DTA methods (66).

The IR data are presented in Fig. 6. The main helix bands of both monomer units overlap: 998 and 841 cm^{-1} for polypropylene, 997 and 848 cm^{-1} for poly–4-methylpentene-1 (see Section III.C.2e). For this reason, the band at 998–997 cm^{-1} was chosen as the measure of the total helix content of the copolymer, the 918 and 973 cm^{-1} bands being the internal standards of the 4-methylpentene-1 and propylene units, respectively. In Fig. 6 the A_{998}/A_{918} and A_{998}/A_{973} ratios for the copolymers are compared with the same ratios for the products of the successive polymerization of both monomers in one run (polyallomers), which from the spectral point of view are equivalent to the homopolymer mixtures. There are also two calculated curves in Fig. 6, the upper one corresponding to these ratios for mixtures of two homopolymers and the lower one for random copolymers ($r_1 r_2 = 1$). The calculations were made with Eqs. (8) and (12) for the following parameters: 998 cm^{-1} (propylene): $n = 10$, $R' = 25$; 973 cm^{-1} (propylene): $n = 4$, $R' = 25$; 997 cm^{-1} (4-methylpentene-1): $n = 5$, $R'' = 25$. The spectral data for the polyallomers coincide with the calculations for the homopolymer mixtures, but the data for the copolymers deviate significantly from the calculations for random copolymers, so demonstrating the tendency toward block formation. This tendency was confirmed by IR data on the 841 cm^{-1} band intensity and by X-ray measurements. The depression of melting points for these copolymers is also very weak (160° C for 85–70 mol.-%, and $\sim 157°$ C for ~ 50 mol.-% of propylene) (66).

It is interesting to note that all these independent results proving the high blockiness of the propylene–4-methylpentene copolymers are not supported by the $r_1 r_2$ value (~ 2).

6. Propylene–Styrene Copolymers

The copolymerization of these two monomers has been effected mainly with highly isospecific catalytic systems like TiCl$_3$—Al(C$_2$H$_5$)$_3$ or VCl$_3$—Al(C$_2$H$_5$)$_3$ (47, 98, 159–162).

The analysis of the copolymer composition is based mainly on IR spectra. For example, the 1603 and 1380 cm^{-1} bands can be used, the former being assigned to the benzene ring vibration and the latter to the methyl group deformation mode (47). The analytical expression was calibrated with the homopolymer absorption coefficients

$$A_{1380}/A_{1603} = 0.235 + 1.80 \, C_{st}/(1 - C_{st}),$$

where C_{st} is the mole fraction of styrene in the copolymer.

The $r_1 r_2$ values predict that the copolymers obtained with the TiCl$_3$—Al(C$_2$H$_5$)$_3$ system have a strong tendency to block formation [$r_1 r_2 \sim 6$, according to (159)]. Experimental data confirm this suggestion. Figure 7 presents the quantitative data for the distribution of styrene and propylene units in these copolymers (47). It is evident from Fig. 7 that both monomers are situated mainly in long sequences to which corresponds the $r_1 r_2$ value ~ 5 (Fig. 7a), which is obviously higher than those of the ethylene–propylene copolymers (Fig. 7b).

The copolymers are crystalline over a wide compositional range (162) and have high melting points (200–150° C) (161). Nevertheless, these products are real copolymers rather than mixtures of two homopolymers. This follows directly from Fig. 7 and was confirmed by the fractionation data (161, 162). The melting-point depression of propylene units in copolymers of low styrene content (159, 161) also supports this conclusion [see Fig. 14 in (159)]. The positions of the styrene unit bands in the IR spectra of the propylene-rich samples is typical for the isolated units.

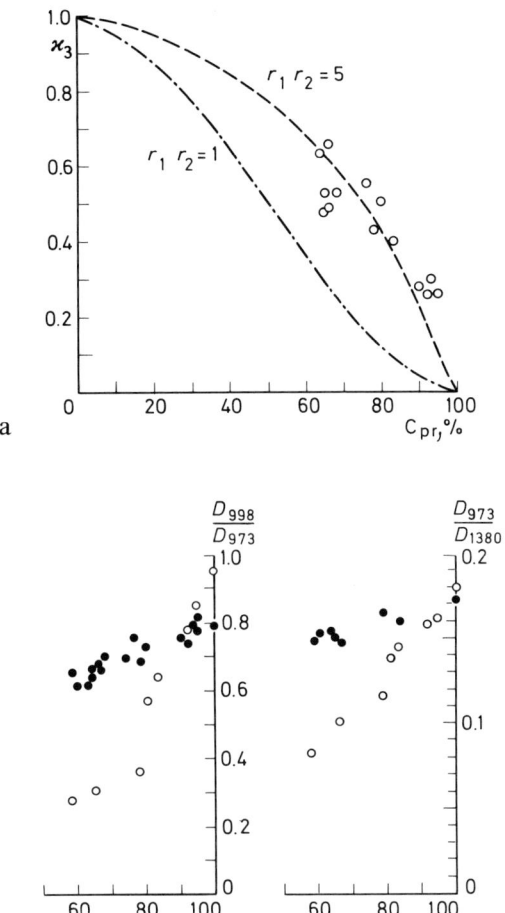

Fig. 7a and b. Distribution of monomer units in propylene–styrene copolymers (*47*). a) Distribution of styrene units. Experimental points are for the 565 cm^{-1} band. Calculations are for $n = 2$, $r_1 r_2 = 1$ and 5. b) Comparison between the A_{998}/A_{973} and A_{973}/A_{1380} ratios (the parameters of the distribution of propylene units in long isotactic sequences, see Table 1) for propylene–styrene copolymers (●) and ethylene–propylene copolymers (○)

The bands are at 1070 (*98, 159*) to 1074 (*47*) cm^{-1} and 556 to 550 cm^{-1} (*47, 98, 159*) (see Table 2).

NMR spectra of these copolymers were also studied at different temperatures (*160*). The aromatic proton signals in the propylene-rich copolymer spectra are singlets at $\sim \tau 2.9$ and with increasing styrene content they shift to a higher field (from $\tau 2.88$ to $\tau 3.02$) due to the shielding effect of the phenyl groups of the neighboring styrene units.

7. Propylene–Vinyl Cyclohexane Copolymers

The copolymers obtained with the catalytic system $TiCl_3$—$Al(C_2H_5)_2Cl$ have been studied by IR, X-ray, and DTA methods (*46, 109*).

The copolymer compositions were determined using the 1380 cm^{-1} band for propylene and the 893 cm^{-1} band for vinyl cyclohexane. The analytical expression is $A_{1380}/A_{893} = 2.14 C_{pr}/(1 - C_{pr})$, where C_{pr} is the mole fraction of propylene in the copolymers. The expression was calibrated with absorption coefficients and was checked by studies on homopolymer mixtures.

These copolymers have a significant tendency to block formation ($r_1 r_2 \sim 4$), as proved by IR data (Fig. 8) demonstrating good correlation between the measured and calculated content of propylene units in isotactic sequences. The melting-point depression for these copolymers is also much less than that for random ones, (*109*) which again confirms the block structure.

8. Other Propylene Copolymers

The copolymerization of propylene with different olefins containing benzene and naphthalene rings (p-N, N-dialkylaminostyrenes, 1- and 2-vinyl naphthalene, allylbenzene, 4-phenylbutene-1) with the $TiCl_3$—$Al(C_2H_5)_3$ system has been studied (*164*). The copolymer composition was measured either chemically or with IR spectra. The extraction data, IR spectra and changes in stability with respect to thermal oxidation confirm the fact of copolymer formation.

C. Butene-1 Copolymers

1. Butene-1–Ethylene Copolymers (see Section IV.A.2)

2. Butene-1–Propylene Copolymers (see Section IV.B.3)

3. Butene-1–Linear Olefin Copolymers

A careful examination has been made of the crystal structure of butene-1–linear olefin copolymers (*145*).

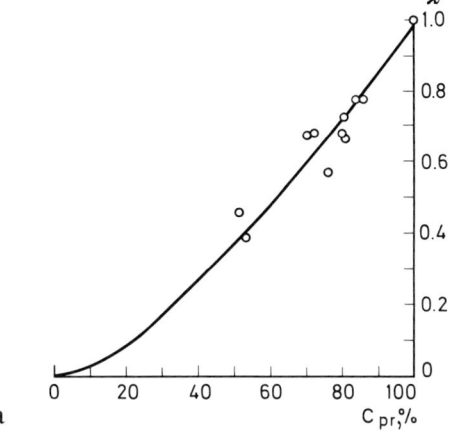

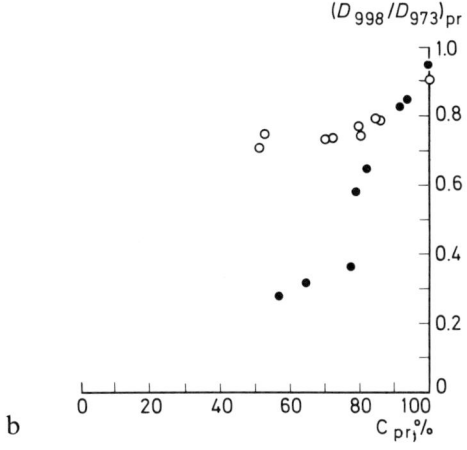

Fig. 8a and b. Distribution of propylene units in propylene–vinyl cyclohexane copolymers (46). a) Data for the 973 cm^{-1} band. α-specially normalized values of A_{973}/A_{1380} – see (11). Calculations are for $n = 4$, $r_1 r_2 = 3.9$. b) Comparison between the A_{998}/A_{973} ratios (the parameters of the distribution of propylene units in long isotactic sequences) in propylene–vinyl cyclohexane copolymers a (○) and in ethylene–propylene copolymers (●)

The simplest situation was found to exist in the case of the highest olefins (nonene to octadecene). In these copolymers the incorporation of quite small quantities of high-olefin units into the copolymers gives rise to a significant decrease in polybutene-1 crystallinity. One of the examples is decene-1:

decene-1 content, %	0	2	5.5	7	12
X-ray crystallinity, %	55	38	27	16	9

These data clearly indicate random copolymer formation. The melting-point depression and the widening of the melting-point ranges also strongly support this conclusion.

In the case of butene-1–pentene-1 copolymers, the crystallinity value hardly changes throughout the whole range of composition (from 65% for polybutene-1 to 39% for the copolymer with 49% C_{but}), nor do the melting points (from 135 to 95°C in the same range). The type of crystallinity is always that of polybutene-1 Modification I, but with spacing variations. All these phenomena are explained as the results of practically ideal isomorphism of both units in the copolymers [$r_1 r_2 = 0.22$ for the VCl_3—$Al(C_2H_5)_3$ system (166)].

The properties of copolymers of butene-1 with hexene-1 and octene-1 are intermediate between those of pentene-1 and the highest olefin copolymers, due to the partial isomorphic behavior in these systems.

An unusual way to produce butene-1 copolymers with other linear olefins has been described (165, 166); these copolymers were obtained starting from β-olefins as a result of a monomer isomerization process. The isomerization of butene-2 and pentene-2 to butene-1 and pentene-1 was confirmed by the presence of the 766 cm^{-1} (ethyl group) and ~740 cm^{-1} (propyl group) bands, and the copolymer compositions were measured as functions of the A_{766}/A_{1380} ratios (166).

4. Butene-1-3-Methylbutene-1 Copolymers

The structure of these copolymers has been studied (39, 42, 145).

The copolymer compositions have been measured mainly by IR methods, for example (42) by using the 746 cm^{-1} band (ethyl group mode) and 1180 cm^{-1} band [isopropyl group mode, insensitive to monomer distribution (53)].

The copolymers are highly crystalline (50–60%) over the whole range of composition (39, 42) and have high melting points (117–310°C). At least six crystalline phases were observable, varying in proportion with the copolymer composition; this fact was attributed to the high degree of isodimorphism and cocrystallization. Nevertheless, some peculiarities of the behavior of copolymer crystallinity suggest a tendency to block formation.

5. Butene-1–4-Methylpentene-1 and Butene-1–4,4-Dimethylpentene-1 Copolymers

X-ray and melting-point studies of these copolymers have been carried out (*145*). All the copolymers studied are crystalline over the whole range of composition (the total crystallinity diminishes from 55–51% for polybutene-1 to $\sim 30\%$ for 1:1–1:2 copolymers, and then rises again to 61–65% for another polyolefin). The melting point variations are also small. True cocrystallization does not occur in these cases owing to the difference in monomer unit sizes. These facts led the authors (*145*) to conclude that there was strong compositional inhomogeneity.

D. 3-Methylbutene-1 Copolymers

1. 3-Methylbutene-1–Ethylene Copolymers (see Section IV.A.5)

2. 3-Methylbutene-1–Butene-1 Copolymers (see Section IV.C.4)

3. 3-Methylbutene–Higher Linear Olefin Copolymers

The structure of 3-methylbutene-1 copolymers with pentene-1 and decene-1 has been studied (*39*). Composition was determined by the IR method. The A_{1050}/A_{1150} ratio was used in the case of 3-methylpentene-1–pentene-1 copolymers, and the analytical curve was calibrated with the homopolymer mixtures.

Decene-1 units in the copolymer chains practically cannot be incorporated onto the poly-3-methylpentene-1 lattice due to the steric hindrance of the pendant groups. These copolymers are nearly random $[r_1 r_2 = 1.73$ for the $TiCl_4$—$Al(C_2H_5)_3$ system] and the crystallinity of the poly-3-methylbutene-1 type virtually vanishes after the introduction of $\sim 40\%$ of decene-1 into the copolymer chain.

Contrary to this, poly-3-methylbutene-1 and polypentene-1 (Modification II) have practically the same crystallographic parameters, and the copolymers of these monomers are crystalline over the total range of composition (the polypentene-1 crystallinity is 74%, that of poly-3-methylbutene-1 is $\sim 90\%$, and the minimum crystallinity of the copolymer with 25% C_{3mbl} is $\sim 30\%$). The copolymer melting points change almost linearly from $\sim 71°$C to $\sim 289°$C in this range. The absence of IR data does not allow any conclusion about the monomer distribution in these copolymers. The only IR data available for 3-methylbutene-1–heptene-1 copolymers (*53*) show a gradual decline in the relative intensities of the 778 and 1218 cm^{-1} bands with increasing heptene-1 content, indicating real copolymer formation.

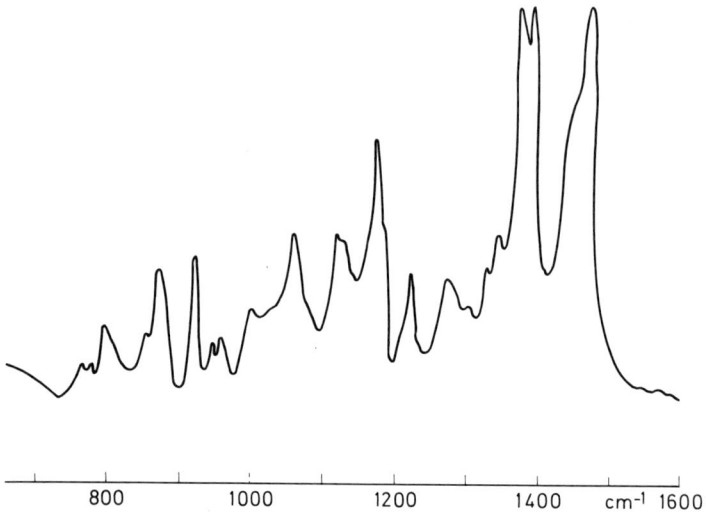

Fig. 9. IR spectrum of a 3-methylbutene-1–4-methylpentene-1 copolymer obtained with the $TiCl_3$—$Al(C_2H_5)_2Cl$ system

4. 3-Methylbutene-1–4-Methylpentene-1 Copolymers

Data on this copolymer structure have been presented (39) and (113). These copolymers are highly crystalline (39) and have very high melting points (39, 113):

4-methylpentene-1 content, %	100	93	87	80	75	52–50	25	0
melting point, °C	248–240	240	225	236	208	230–218	260	300

Nevertheless, the IR spectrum of such a copolymer (Fig. 9) demonstrates the absence of helix bands in both monomers, indicating random unit distribution, in agreement with the $r_1 r_2$ value ~ 1 (39). The high crystallinity of the copolymer is due to isomorphism (39).

5. 3-Methylbutene-1–Vinyl Cyclohexane Copolymers

The composition of these copolymers is determined in the same way as for 4-methylpentene-1–vinyl cyclohexane copolymers (Section IV.E.7).

The IR spectra demonstrate that real copolymerization of these monomers takes place when $TiCl_4$—$Al(isoC_4H_9)_3$ is used as the catalyst system. The 778 cm^{-1} band [the helix band of poly-3-methylbutene-1, see Section II.C.2e] disappears from the copolymer spectra, beginning from $\sim 50\%$ vinyl cyclohexane content, and the relative intensity of the 1120 cm^{-1} band decreases significantly:

3-methylbutene-1 content, mol.-%	100	90	84	55	40
$(A_{1120}/A_{1385})_{cop}/(A_{1120}/A_{1385})_{p3mb1}$	1.0	0.72	0.72	0.22	0.14

E. 4-Methylpentene-1 Copolymers

1. 4-Methylpentene-1–Ethylene Copolymers (see Section IV.A.6)
2. 4-Methylpentene-1–Propylene Copolymers (see Section IV.B.5)
3. 4-Methylpentene-1–Butene-1 Copolymers (see Section IV.C.5)
4. 4-Methylpentene-1–Higher Linear Olefins Copolymers

Data on the structure of the copolymers of 4-methylpentene-1 with different linear olefins (pentene-1, hexene-1, octene-1, decene-1 and octadecene-1) have been presented (48, 114, 167).

a) 4-Methylpentene-1–Pentene-1 Copolymers. The products of the copolymerization of these monomer with a $TiCl_3$-containing system are crystalline over the whole range of composition (114):

4-methylpentene-1, %	100	89	66	55	33	17	0
crystallinity, %	68	59	45	40	43	47	50

and have high melting points between those of the two homopolymers (245 and 80° C). The formation of a crystalline product is due to isodimorphism (114). The absence of homopolymers in the copolymerization products and their X-ray and melting point data is consistent with a significant degree of randomness in this copolymerization "but show that the copolymers are not homogeneous and ... contain chains with a range of comonomer composition and/or degrees of blocking" (114).

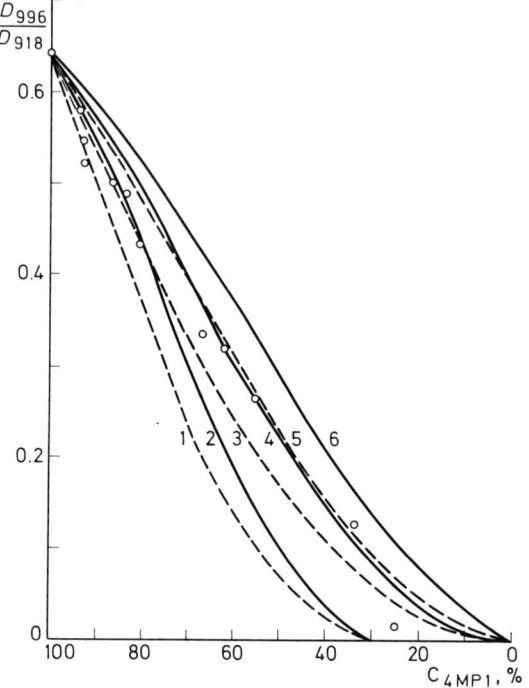

Fig. 10. Distribution of 4-methylpentene-1 units in 4-methylpentene-1–hexene-1 copolymers obtained with the TiCl$_3$—Al(C$_2$H$_5$)$_2$Cl system. Calculations are carried out for $r_1 r_2 = 1$, $n = 5$ (1); $r_1 r_2 = 1$, $n = 4$ (2); $r_1 r_2 = 3$, $n = 5$ (3); $r_1 r_2 = 3$, $n = 4$ (4); $r_1 r_2 = 5$, $n = 5$ (5); $r_1 r_2 = 5$, $n = 4$ (6)

b) 4-Methylpentene-1–Hexene-1 Copolymers. Analysis of copolymer composition has been performed by IR methods using the 726 cm^{-1} band for the hexene-1 content and the 918 or 1170 cm^{-1} bands for the 4-methylpentene-1 content (48, 167), and the 4310 cm^{-1} band as the sample thickness standard (114). The calibration curves are:

$$A_{918}/A_{726} = 0.423\,C_{4mp1}/(1 - C_{4mp1}) + 0.18 \qquad (48)$$

$$A_{726}/A_{1170} = 0.934(1/C_{4mp1}) - 0.934 \qquad (48)$$

$$A_{728}/A_{1168} = 0.1 + 1.22\,C_{hex}/C_{4mp1} \qquad (167).$$

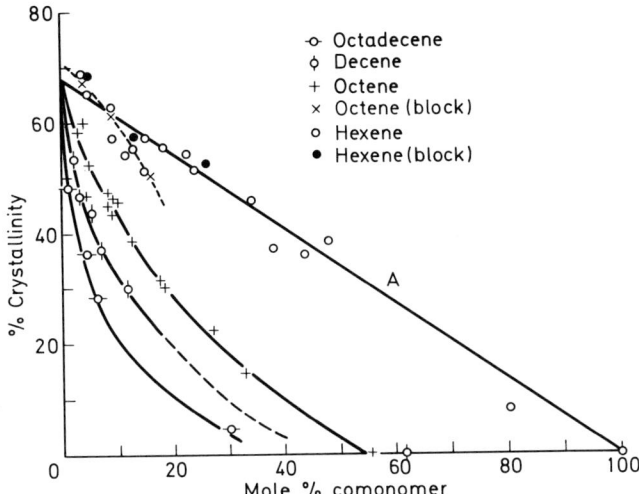

Fig. 11. Crystallinity of the copolymers of 4-methylpentene-1 with different linear olefins
Reproduced from (114) by permission of the publishers, IPC Science & Technology Press Ltd, Guildford, U.K.

(C_{4mp1} is the mole fraction of 4-methylpentene-1 in the copolymers; calibration with absorption coefficients and with homopolymer mixtures).

The unit distribution was estimated by IR and melting-point methods. IR data for the 4-methylpentene-1 units are based on the relative intensity of the 997 cm^{-1} band in the copolymer spectra (Fig. 10) and show that these copolymers have a significant tendency to block formation ($r_1 r_2 \sim 3$–5). This conclusion was supported by melting-point measurements (114, 167). The copolymer melting points are evidently lower than those for homopolymer mixtures, thus demonstrating that a real copolymerization takes place (167), but they have an upward deviation from the theoretical curve calculated for the random copolymer model by means of the Flory equation (Section III.E) with $\Delta H = 4710$ kal/mol (171).

The crystallinity measurements are of little significance in this case due to the practically perfect isomorphism (114, 167). Both homopolymers have 7/2 helices, and hexene-1 readily replaces 4-methylpentene-1 units in the lattice (114). This fact explains linear dependence of the composition on crystallinity (Fig. 11).

c) **Copolymers of 4-Methylpentene-1 with Octene-1, Decene-1 and Octadecene-1.** The octadecene-1 units are not incorporated into the lattice of poly-4-methylpentene-1 at all (no *a*-axis expansion was found) and this monomer pair is thus free from any possible influence of isomorphism on copolymer crystallinity (*114*). As Fig. 11 demonstrates, the crystallinity of these copolymerization products decreases sharply in this case. The low residual crystallinity of copolymers with $\sim 30\%$ of octadecene-1 and the melting-point data (higher than expected for random copolymerization according to the Flory equation) suggest a distinct deviation of these copolymers from randomness, *i.e.* a tendency to form blocks.

The octene-1 and decene-1 units alter the poly-4-methylpentene-1 lattice to a some extent. The X-ray (Fig. 11) and melting-point data demonstrate and evident tendency to block formation in these cases also.

5. 4-Methylpentene-1–3-Methylbutene-1 Copolymers
(see Section IV.D.4)

6. 4-Methylpentene-1–Styrene Copolymers

Some published studies of these copolymer structures (*45, 168–170*) are contradictory. The IR data demonstrate (*45*) the random structure of the copolymers with the catalytic system $TiCl_3$—$Al(isoC_4H_9)_3$. The distribution of styrene units in long isotactic blocks was measured with the 1084 cm^{-1} band absorbance (Fig. 12) by a method similar to that described in Ref. (*65*) and with the 565 cm^{-1} band absorbance [see Fig. 6 in (*45*)], as well as data on 4-methylpentene-1 unit distribution [Fig. 8 in (*45*)]. All are in agreement with the $r_1 r_2$ value of magnitude 1 rather than with the values 3.3 to 3.8 that follow from the f–F correlations (*45, 169*). It is evident, for instance, from the IR spectra of these copolymers [Fig. 4b in (*45*)] that the great majority of the styrene units are isolated (555 and 1073 cm^{-1} bands) in the copolymers having less than 50% of styrene.

On the other hand the X-ray diffractometric scans (*168*) show clear features of semicrystalline products. The X-ray data in (*45*) confirm this, but some distinct differences were found between the copolymer and homopolymer spectra, suggesting a possibility of isomorphism in this system. A more elaborate X-ray study is probably needed to explain the divergence between the IR and X-ray estimates of copolymer blockiness.

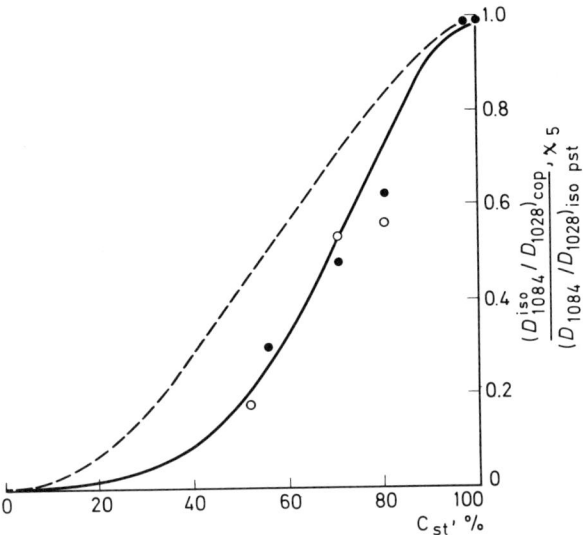

Fig. 12. Distribution of styrene units in 4-methylpentene-1–styrene copolymers obtained with the catalytic systems $TiCl_3$—$Al(isoC_4H_9)_3$ (●) and $TiCl_3$—$Al(C_2H_5)_3$ (○). Experimental points are for the 1084 cm^{-1} band (see Table 2). Calculations are carried out for $n = 4$, $r_1 r_2 = 1$ (solid line) and $n = 4$, $r_1 r_2 = 3.5$ (dotted line) Reproduced from (45) by permission of European Polymer Journal. Pergamon Press Ltd.

7. 4-Methylpentene-1–Vinyl Cyclohexane Copolymers

The copolymer composition measurements were done by IR (A_{1388}/A_{1451}, calibration with homopolymer mixtures) and chromotographic methods (43). The monomer distribution was estimated by means of the 997 cm^{-1} band method, as used for 4-methylpentene-1 copolymers with ethylene (Section IV.A.6), hexene-1 (Section IV.E.4b), and styrene (Section IV.E.6). The $(A_{997}/A_{918})_{cop}$ versus C_{4mp1} dependence, measured from the IR spectra, satisfactorily correlated with Eq. (8) for $r_1 r_2 \sim 1$ [see Fig. 4 in (43)]. The evaluation of vinyl cyclohexane unit distribution was also studied.

8. Other 4-Methylpentene-1 Copolymers

Few copolymers of 4-methylpentene-1 with branched olefins have been studied. The isomorphism phenomenon was found in 4-methylpentene-1–racemic 4-methylhexene-1 copolymers, the a-axis spacing and melting points being in linear dependence on the copolymer composition over the whole range (172). 4-Methylpentene-1–3-methylpentene-1 copolymers (173) are also crystalline over the whole compositional range and have high melting points (235–275° C) due to isomorphism.

F. Styrene Copolymers

1. Styrene–Ethylene Copolymers (see Section IV.A.7)
2. Styrene–Propylene Copolymers (see Section IV.B.6)
3. Styrene–Higher Olefin Copolymers

Very few data are available on the structure of the copolymers of styrene with pentene-1, hexene-1 and heptene-1 obtained with the catalytic systems $TiCl_3$—$Al(C_2H_5)_3$ and $TiCl_4$—$Al(C_2H_5)_3$ (170, 174). The position of the styrene-ring bands in the spectra of styrene–pentene-1 copolymers containing 2.5–6.7% of styrene (~ 550 cm^{-1}) shows that in these compositions the styrene units are mainly isolated in the copolymer chains (174).

4. Styrene–4-Methylpentene-1 Copolymers (see Section IV.E.6)
5. Styrene–Vinyl Cyclohexane Copolymers

The copolymer compositions were determined by IR spectra (44). The analytical bands at 1603 cm^{-1}, 1494 cm^{-1} (benzene ring modes), and 1455–1451 cm^{-1} (methylene deformation modes in styrene and vinyl cyclohexane rings) were chosen and the calibration dependences were obtained with absorption coefficients and were checked by recording the IR spectra of homopolymer mixtures:

for low styrene content $A_{1451}/A_{1603} = 1.92 + 3.05(1 - C_{st})/C_{st}$,

for high styrene content $A_{1455}/A_{1603} = 2.53 + 2.65(1 - C_{st})/C_{st}$.

The distribution of styrene units was examined with the relative intensities of the 1084 and 565 cm^{-1} bands (see Table 2). Figure 13 shows that these spectral data are in satisfactory agreement with the statistical calculations for $r_1 r_2 \sim 0.3$ (the $r_1 r_2$ value from the f–F correlation is 0.37) (44).

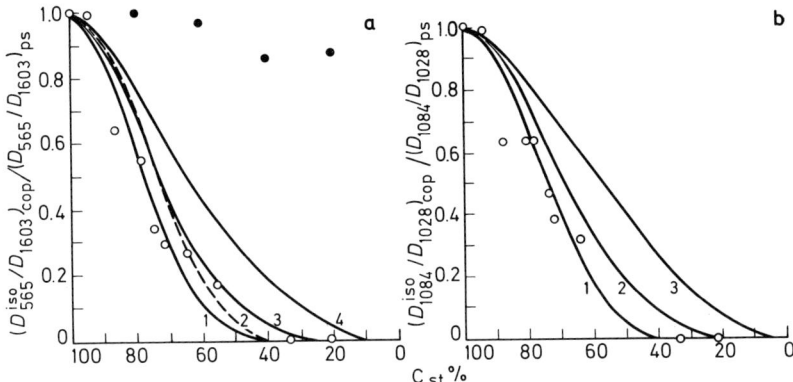

Fig. 13a and b. Distribution of styrene units in styrene–vinyl cyclohexane copolymers. a) Experimental data are for the 565 cm^{-1} band (see Table 2). Calculations are carried out for the cases $r_1 r_2 = 0.3, n = 5$ (1); $r_1 r_2 = 0.3, n = 4$ (2); $r_1 r_2 = 1$, $n = 5$ (3); and $r_1 r_2 = 3, n = 5$ (4). b) Experimental data are for the 1084 cm^{-1} band (see Table 2). Calculations are carried out for the values $r_1 r_2 = 0.3$, (1), 1 (2), 3 (3), and $n = 4$. ● = homopolymer mixtures, ○ = copolymers
Reproduced from (44) by permission of European Polymer Journal. Pergamon Press Ltd.

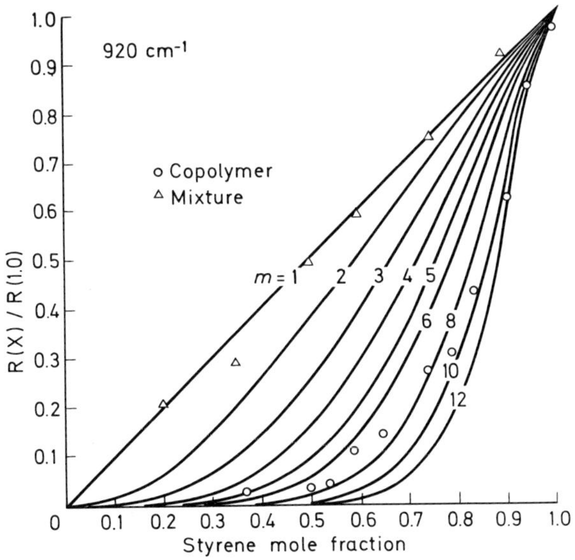

Fig. 14. Evaluation of the threshold parameter n for the 920 cm^{-1} band in the IR spectra of styrene–styrene-α-D$_1$ copolymers
Reproduced from (65) by permission of Hüthig & Wepf Verlag

6. Styrene–Styrene-α-D_1 Copolymers

These copolymers were used as a model for the styrene-containing product of constant conformational structure and crystallinity, and with varying styrene content to study the lengths of isotactic styrene sequences sufficient of the appearance of different helix bands in the IR spectra (65). Some of these results are presented in Table 2. Figure 14 demonstrates one example of such a sequence determination for the 920 cm^{-1} band. The statistical curves there are $\varkappa_{n+1} \cdot C_{st}$ products calculated with Eqs. (11) and (12) for $r_1 r_2 = 1$ and different n.

7. Other Styrene Copolymers

Some data have been published on the copolymers of styrene with branched olefins (169, 170) and with vinyl aromatic compounds (175). Styrene–5-methylhexene-1 (169) and styrene–4-methylhexene-1 (170) copolymers are crystalline over the whole compositional range. Nevertheless, these products are real copolymers as is clear from the solubility data and melting-point depression (169). There are three reasons for the crystallinity of these products: high $r_1 r_2$ values (2.3–2.4), significant compositional inhomogeneity of the samples, and isomorphism of sterically similar units.

Some of the styrene–vinyl aromatic compound copolymers also demonstrate isomorphism (175). The best examples are styrene–o-F-styrene copolymers, which are highly crystalline products with melting points in the 235–270° C range. The helix bands in the IR spectra of these copolymers disappear, beginning from the 30% content of another comonomer, confirming random copolymerization (175).

G. Optically Active Copolymers

There are some data on copolymers containing optically active centers in pendant groups (50, 177, 178). The copolymerization of (s)-4-methylhexene-1 with 4-methylpentene-1 (50) produces copolymers with a disproportionally high optical activity compared with that of mixtures of two homopolymers of the same composition. One possible explanation for this fact is that left-hand helix formation of the polymer chain may be facilitated by the optically active units in it, so allowing the asymmetric carbon atoms in the main chain to manifest their optical activity.

When racemic 3,7-dimethyloctene-1 and 3-methylpentene-1 were polymerized respectively with (s)-3-methylpentene-1 and (r)-3,7-dimethyloctene-1 (*177*), the optical activity and the IR analysis of the copolymer fractions demonstrated that copolymerization takes place predominantly between the optically active monomer and the monomer in the racemic mixture having the same chirality, the other antipode giving the homopolymer. Copolymer formation in these cases was detected by the IR method, using the ratio of the absorbances of the 763 cm^{-1} band (ethyl group mode in 3-methylpentene-1 units) and those of the 732 cm^{-1} band [(CH$_2$)$_3$ rocking mode of the 3,7-dimethyloctene-1 units] as a qualitative measure of the copolymer composition.

V. Compositional Inhomogeneity of Olefin Copolymers

Practically every paper dealing with the copolymerization of olefins contains data about polymer fractionation (mainly with different solvents). These data show that in the majority of cases different copolymer fractions have different compositions.

There are three possible explanations for copolymer inhomogeneity in respect to composition
1) Even in the ideal case, different copolymer molecules have different composition for statistical reasons. Where reasonably high molecular products are concerned, this distribution is narrow enough to be neglected.
2) One of the most obvious reasons for copolymer inhomogeneity is variation of the feed ratio F during a run [see *e.g.* Fig. 8 in (*156*)]. In such cases the possible widening of composition can be evaluated from an analysis of the monomer mixture at different stages of a run and with r_1 and r_2 values.
3) Even in experiments performed at strictly constant F, the copolymers obtained with heterogeneous catalysts sometimes have a significant compositional distribution (*32, 129, 138, 145, 179, 180*). The possible explanation is that different active sites on the catalyst surface have different activities in respect to comonomers, *i.e.* different r_1 and r_2 values (*66, 145, 169*). On the other hand, when soluble catalytic systems are used, a relatively narrow compositional distribution is observed (*32, 179, 180*) showing one type of active site.

Two questions arise in connection with the compositional inhomogeneity of the copolymer samples: What is the meaning of r_1

and r_2 values in the case of the multiple-site mechanism, and what is the correlation between the effective $r_1 r_2$ values and the real monomer distribution in the copolymer chains? Is the tendency to block formation which is typical for some olefin copolymers (see table for $r_1 r_2$ products) the real phenomenon or is it rather the result of the wide compositional distribution? (For example the equimolar mixture of polyethylene and polypropylene can be formally regarded as a "copolymer" with composition 1:1 and with $r_1 r_2 \to \infty$).

A. r_1 and r_2 Values in the Multiple-site Model

It is clear that r_1 and r_2 values in the mono-site model have great significance: each is the ratio of two propagation constants and their product determines all sequence distribution functions (see Section II, where all equations presented are for the mono-site model). Is this also true of the multiple-site model, or are the r_1 and r_2 parameters here simply a useful mathematical device for correlating two experimental values, F and f (138)?

Some calculations were performed to elucidate this problem (129, 181). The case of the mixture of two copolymers was examined (181) for two separate variants

A) A mixture of two copolymers (their proportions are p_1 and $1 - p_1$) having the same composition but characterized by different $r_1 r_2$ values. In this case, one of the monomer distribution functions [the fraction of isolated units, δ_1 of Eq. (10)] is determined by some effective $(r_1 r_2)_{\text{eff}}$ value intermediate between $(r_1 r_2)_1$ and $(r_1 r_2)_2$. For copolymers of composition near 100% the $(r_1 r_2)_{\text{eff}}$ value is close to $p_1 (r_1 r_2)_1 + (1 - p_1)(r_1 r_2)_2$, and for the copolymers of intermediate composition $\lg(r_1 r_2)_{\text{eff}} \simeq p_1 \lg(r_1 r_2)_1 + (1 - p_1) \lg(r_1 r_2)_2$.

B) A mixture of two copolymers of different composition but characterized by the same $r_1 r_2$ value, i.e. $(r_1 r_2)_1 = (r_1 r_2)_2$. In this case the situation strongly depends on the p_1 value as well as on the difference between the composition of the copolymers. As this difference increases, the $(r_1 r_2)_{\text{eff}}$ value increases sharply [$\lim(r_1 r_2)_{\text{eff}} \to \infty$ for the mixture of two homopolymers] and in many cases $(r_1 r_2)_{\text{eff}}$ is higher than $(r_1 r_2)_1$ [see also (129)].

The case of the multiple-site model has been discussed (129) (N types of sites) and a generalized expression was given for the copolymer composition. The two main conclusions of this paper are that the $(r_1 r_2)_{\text{eff}}$ value taken from the mean composition measurements falls between $r_1 r_2$ values of the individual species, and that when $(r_1 r_2)_{\text{eff}}$ and average copolymer composition are used to calculate an average propagation probability [Eq. (9)], the number of units in long sequences is underestimated, i.e. $(r_1 r_2)_{\text{eff, unit distribution}} > (r_1 r_2)_{\text{eff, copolymer composition}}$.

We examined a model of the continuous distribution of active sites in respect to r_1 and r_2 parameters. From our earlier work (180) we assumed the existence of close dependence between the stereospecificity of the active sites and their r_1 and r_2 values. The main conclusions that follow from our calculations are:

1) The resulting $\bar{f} - F$ dependence can be described with reasonable accuracy by two parameters $(r_1)_{\text{eff}}$ and $(r_2)_{\text{eff}}$.

2) The "real" blockiness of the modeled copolymers is always higher than that calculated with $(r_1 r_2)_{\text{eff}}$. Hence, we underestimate the blockiness of the copolymers obtained with a multiple-site model if we use $(r_1 r_2)_{\text{eff}}$ as the measure of copolymer randomness [the same conclusion as in (129)].

3) The degree of this underestimation is of the order of 1–2 units in the $r_1 r_2$ scale (for example, instead of $r_1 r_2 = 4.5$ following from the $\bar{f} - F$ correlation, the true $r_1 r_2$ value describing the monomer distribution is 5–6). Moreover, it increases with decreasing catalyst stereospecificity. This conclusion correlates qualitatively with the experimental data: for highly isospecific systems the distribution is relatively narrow [see Fig. 8 in (156) where the composition range is ~10%] but increases for less specific systems (180, 182).

B. Structure of Copolymer Fractions

It is evident from the previous section that in the multiple-site model the copolymer is a mixture of copolymers of different composition; this fact alone accounts to a some extent for the experimentally determined copolymer blockiness. However, a study of the copolymer fractions demonstrates that it cannot explain all the results. The blockiness of the copolymer fractions remains high; therefore the tendency to block formation is an intrinsic property of active sites rather than merely the result of copolymer mixing. For example, when the propylene–butene-1 copolymer was fractionated (156), the resulting narrow fractions of composition $C_{\text{pr}} \sim 30$–50% exhibit X-ray diffraction curves containing crystallinity peaks of both polypropylene and polybutene-1 type. An analogous result was demonstrated by us (66): the blockiness of the fractions of propylene–4-methylpentene-1 copolymers insoluble

Table 5. Reactivity ratios for high olefin copolymerization

Olefin pair	Catalytic system	r_1	r_2	$r_1 r_2$	Ref.
Ethylene–butene-1	$TiCl_3$-containing catalytic system	60.0	0.025	1.5	(138)
	$TiCl_3$—$CH_3 TiCl_3$	3.6	0.16	0.58	(183)
	VCl_3—$Al(C_6H_{13})_3$	26.96	0.043	1.16	(184)
	VCl_4—$Al(C_6H_{13})_3$	29.6	0.019	0.56	(184)
	VCl_4—$Al(isoC_4H_9)_2Cl$	32.5	0.0186	0.60	(35)
	V acetylacetonate–$Al(isoC_4H_9)_2Cl$	26	0.022	0.57	(35)
	$(C_5H_5)_2 TiCl_2$—$Al(C_2H_5)_2Cl$, 20° C, in toluene	18.4	0	0	(144)
	20° C, ethyl chloride	15.5	0	0	(144)
Ethylene–pentene-1	VCl_4—$Al(isoC_4H_9)_2Cl$	42.1	0.015	0.63	(35, 151)
	V acetylacetonate–$Al(isoC_4H_9)_2Cl$	32.2	0.014	0.45	(35, 151)
Ethylene–hexene-1	V acetylacetonate–$Al(isoC_4H_9)_2Cl$	33.2	0.0145	0.48	(149)
Ethylene–3-methylbutene-1	V acetylacetonate–$Al(isoC_4H_9)_2Cl$	46	0.0125	0.58	(163)
	V acetylacetonate–$Al(isoC_4H_9)_2Cl$	243.0	0.001	0.24	(149, 152)
Ethylene–4-methylpentene-1	$TiCl_3$—$Al(C_2H_5)_2Cl$	195.1	0.028	4.8	(153)
	V acetylacetonate–$Al(isoC_4H_9)_2Cl$	47	0.015	0.70	(153)
Ethylene–styrene	$TiCl_3$—$Al(C_2H_5)_3$	81	0.012	0.97	(194)
	VCl_3, $VOCl_3$, VCl_4—$Al(C_2H_5)_3$	81	0.010	0.81	(195)

Table 5 (continued)

Olefin pair	Catalytic system	r_1	r_2	$r_1 r_2$	Ref.
	$TiCl_3$-containing catalytic system	4.67	0.511	2.39	(138)
	$TiCl_3$—$Al(C_2H_5)_2Cl$-hexamethylphosphoric triamide, 30° C	3.3	0.45	1.48	(156)
	$TiCl_3$—$Al(C_2H_5)_2Cl$-hexamethylphosphoric triamide + H_2, 30° C	4.3	0.8	3.44	(156)
Propylene–butene-1	$TiCl_3$—$Al(C_2H_5)_3$, 60° C	1.62	0.50	0.81	(38)
	$TiCl_3$—$AlC_2H_5)_2Cl$	4.5	0.2	0.90	(185)
	$TiCl_4$—$Al(C_2H_5)_3$, Al:Ti = 2	2.4	0.5	1.20	(186)
	Al:Ti = 5	1.8	0.4	0.72	(186)
	VCl_4—$Al(C_6H_{13})_3$	4.39	0.227	1.00	(184)
	VCl_3—$Al(C_6H_{13})_3$	4.04	0.252	1.02	(184)
	VCl_4-anisole-$Al(C_2H_5)_2Cl$, −78° C	0.7	0.7	0.49	(126)
Propylene–4-methylpentene-1	$TiCl_3$—$Al(C_2H_5)_2Cl$, 70° C	6.5	0.3	1.95	(66)
	$TiCl_3$—$Al(C_2H_5)_3$	20.5	0.30	6.15	(159)
	$TiCl_3$—$Al(isoC_4H_9)_3$	20.0	0.20	4.00	(159)
Propylene–styrene	$TiCl_3$—$Al(C_2H_5)_3$	7.7	0.12	0.92	(187)
	VCl_3—$Al(C_2H_5)_3$	7.2	0.16	1.15	(159)
Propylene–vinyl cyclohexane	$TiCl_3$—$Al(C_2H_5)_2Cl$, 70° C	80	0.049	.392	(109)
Butene-1–pentene-1	VCl_3—$Al(C_2H_5)_3$	0.30	0.74	0.22	(166)
Butene-1–decene-1	$TiCl_3$—$Al(C_2H_5)_2Cl$, 23° C	1.5	0.7	1.1	(185)
Butene-1–3-methylbutene-1	$TiCl_3$—$Al(isoC_4H_9)_3$, 60° C	8.5	0.013	0.11	(188)
	$TiCl_4$—$Al(C_2H_5)_3$, 45° C	6.22	0.33	2.05	(39)

Table 5 (continued)

Olefin pair	Catalytic system	r_1	r_2	$r_1 r_2$	Ref.
3-methylbutene-1–pentene-1	TiCl$_3$—Al(C$_2$H$_5$)$_3$, 45° C	0.28	8.32	2.33	(39)
	TiCl$_4$—Al(C$_2$H$_5$)$_3$, 45° C			1.72	(39)
3-methylbutene-1–decene-1	TiCl$_4$—Al(C$_2$H$_5$)$_3$, 45° C	0.05	34.5	1.72	(39)
3-methylbutene-1–4-methylpentene-1	TiCl$_4$—Al(C$_2$H$_5$)$_3$, 45° C	0.11	9.00	0.99	(39)
3-methylbutene-1–vinyl cyclohexane	TiCl$_4$—Al(iso-C$_4$H$_9$)$_3$	1.02	0.98	1.00	(44)
4-methylpentene-1–hexene-1	TiCl$_3$—Al(C$_2$H$_5$)$_2$Cl	0.62	5.4	3.35	(167)
	TiCl$_3$—Al(C$_2$H$_5$)$_2$Cl	0.37	3.4	1.26	(48)
4-methylpentene-1–3-methylpentene-1	TiCl$_3$—Al(C$_2$H$_5$)$_3$, 80° C	6.2	0.1	0.62	(173)
4-methylpentene-1–styrene	TiCl$_3$—Al(C$_2$H$_5$)$_3$, 45° C	3.67	0.89	3.27	(169)
	TiCl$_3$—Al(iso-C$_4$H$_9$)$_3$, 70° C	3.92	0.98	3.84	(45)
4-methylpentene-1–vinyl cyclohexane	TiCl$_3$—Al(iso-C$_4$H$_9$)$_3$	3.17	0.38	1.20	(43)
Hexene-1–decene-1	TiCl$_3$—Al(C$_2$H$_5$)$_2$Cl, 23° C	1.3	0.9	1.17	(185)
Styrene–hexene-1	TiCl$_3$—Al(C$_2$H$_5$)$_3$, 35° C	0.19	9.75	1.85	(170)
Styrene–heptene-1	TiCl$_3$—Al(C$_2$H$_5$)$_3$, 35° C	0.61	5.70	3.48	(170)
Styrene–4-methylhexene-1	TiCl$_3$—Al(C$_2$H$_5$)$_3$, 35° C	1.80	1.30	2.34	(170)
Styrene–5-methylhexene-1	TiCl$_3$—Al(C$_2$H$_5$)$_3$, 35° C	0.591	4.00	2.36	(169)
Styrene–vinyl cyclohexane	TiCl$_3$—Al(iso-C$_4$H$_9$)$_3$, 60° C	2.12	0.18	0.38	(44)
Styrene–p-methylstyrene		0.82	1.15	0.93	(176)
Styrene–m-methylstyrene	TiCl$_3$—Al(C$_2$H$_5$)$_3$	2.0	0.5	1.00	(176)
Styrene–p-ethylstyrene	TiCl$_4$—Al(C$_2$H$_5$)$_3$	1.0	1.0	1.00	(176)
Styrene–p-F-styrene	TiCl$_3$—Be(C$_2$H$_5$)$_2$	1.5	0.7	1.05	(176)
Styrene–p-Cl-styrene		2.2	0.5	1.10	(176)
Styrene–p-Br-styrene		1.7	0.5	0.85	(176)

in hot heptane is higher than in unfractionated copolymers of similar composition.

Nevertheless, even in the cases when high copolymer blockiness was found, no separate homopolymerization of comonomers took place (37, 156). This means that the homopolymerization and copolymerization processes occur on the same active sites, *i.e.* the sites of the stereospecific olefin polymerization.

VI. Activities of Different High Olefins in Homopolymerization and Copolymerization Reactions

A. Reactivity Ratios for High Olefin Copolymerization

The reactivity ratios for different high olefin pairs are listed in Table 5. This Table does not contain the numerous data for ethylene–propylene copolymers [see (2, 127–129)].

B. Relative Activities of Olefins in Homopolymerization and Copolymerization Reactions

Table 6 contains data that allow the semiquantitative comparison of olefin activities in homopolymerization and copolymerization reactions. The diagonal of Table 6 contains the data on the absolute constants of chain propagation in homopolymerization. When comparing these results, one has to remember that they were obtained with different samples of $TiCl_3$ and under different experimental conditions, and that this introduces some bias in the final results.

Table 6 demonstrates that a good correlation exists between the propagation rate constants for different monomers and their relative activities in the copolymerization reactions represented by r_1 and r_2 values. The order of the olefin activities in the case of heterogeneous complex catalysts according to Tables 5 and 6 is:

ethylene > propylene > olefins with linear chains >
> 4-methylpentene-1 > styrene >
> 3-methylbutene-1 $\simeq$ vinylcyclohexane .

Table 6

	Ethylene	Propylene	Butene-1	Other linear olefins
Ethylene	$K_p = 78$ l/m·s TiCl$_3$(Al)— —Al(C$_2$H$_5$)$_2$Cl, 60° C, (*191*)	15.7–0.11 (*121*) TiCl$_3$—Al(C$_6$H$_{13}$)$_3$	60.0–0.025 TiCl$_3$ (*138*)	33.2–0.0145[b] (pentene-1) (*149*)
Propylene		$K_p = 18$–6.2 l/m·s, 60° C, (*191, 192*) $K_p = 10$–6 l/m·s, 70° C, (*23, 109*) (TiCl$_3$, TiCl$_3$(Al))	4.67–0.51 (*138*) 4.5–0.2 (*185*) 1.62–0.5 (*38*) TiCl$_3$—AlC$_2$H$_5$)$_3$— —Al(C$_2$H$_5$)$_2$Cl	
Butene-1			$K_p = 7.3$ l/m·s (*191*) TiCl$_3$(Al)— —Al(C$_2$H$_5$)$_2$Cl, 60° C	
Other linear olefins				
4-Methylpentene-1				
Styrene				
3-Methylbutene-1				
Vinylcyclohexane				

In the case of olefins with branched alkyl groups the order of activities is:

5-methylhexene-1 ≈ 4-methylpentene-1 > 4-methylhexene-1 >
> 3-methylpentene-1 ≈ 3-methylbutene-1 .

These activity scales are in agreement with previously published data (*127, 176, 193*).

A comparison of these scales with Taft's induction constants σ^* and Taft's steric constants E_s for different alkyl groups leads to the conclusion that both electronic and steric factors influence the double-bond activity in stereospecific catalysis; but that steric factors seem to be more important. An especially significant decrease in the olefin activity was found for olefins branched vicinally to the double bond (3-methylpentene-1, 3-methylbutene-1, vinylcyclohexane, styrene). This is probably connected with the space limitations for monomer coordination in the stereospecific active sites (*15, 16*).

Table 6 (continued)

4-Methylpentene-1	Styrene	3-Methylburene-1	Vinylcyclohexane	
195.1–0.025 TiCl$_3$—Al(C$_2$H$_5$)$_2$Cl	81–0.012 TiCl$_3$—Al(C$_2$H$_5$)$_3$ (194)	243.0–0.001 [b] (149, 152)		Ethylene
6.4–0.31 (66) TiCl$_3$(Al)— —Al(C$_2$H$_5$)$_2$Cl	20.5–0.3 (159) TiCl$_3$—Al(C$_2$H$_5$)$_3$		80–0.049 (106) TiCl$_3$(Al)— —Al(C$_2$H$_5$)$_2$Cl	Propylene
		8.5–0.013 (188) TiCl$_3$—Al(isoC$_4$H$_9$)$_3$		Butene-1
5.4–0.62 (167) 3.4–0.37 (48) (hexene-1) TiCl$_3$—Al(C$_2$H$_5$)$_2$Cl	9.7–0.19 (170) (hexene-1) 5.7–0.61 (170) (heptene-1)	8.32–0.28 (39) (pentene-1) 34.5–0.05 (39) (decene-1) TiCl$_3$, TiCl$_4$		Other linear olefins
K_p = 2.5 l/m · s (190) TiCl$_3$(Al)— —Al(C$_2$H$_5$)$_2$Cl, 70° C	3.92–0.98 (45) 3.67–0.89 (169) TiCl$_3$—Al(C$_2$H$_5$)$_3$— —Al(isoC$_4$H$_9$)$_3$	9.0–0.11 (39) TiCl$_4$—Al(C$_2$H$_5$)$_3$	3.17–0.38 (43) TiCl$_3$—Al(isoC$_4$H$_9$)$_3$	4-Methylpentene-1
	K_p ~ 0.1 l/m · s (189) TiCl$_3$—Al(C$_2$H$_5$)$_3$ 70° C		2.12–0.18 (44) TiCl$_3$—Al(isoC$_4$H$_9$)$_3$	Styrene
			1.02–0.98 (44) TiCl$_4$—Al(isoC$_4$H$_9$)$_3$	3-Methylbutene-1
			K_p = 0.022 l/m · s (109) TiCl$_3$(Al)— —Al(C$_2$H$_5$)$_2$Cl, 70° C	Vinylcyclohexane

[a] K_p are the propagation constants (l/mol · sec) for the corresponding monomer at 60–70° C for TiCl$_3$-containing systems. TiCl$_3$ is the product of the reduction of TiCl$_4$ with H$_2$, Ti or Si (α-type of TiCl$_3$). TiCl$_3$(Al) is the product of the reduction of TiCl$_4$ with Al or AlR$_3$ (δ-type of TiCl$_3$).
[b] These data refer to the V acetylacetonate-Al(isoC$_4$H$_9$)$_2$Cl system. No data are available for TiCl$_3$-containing systems.

VII. Conclusion

The conclusions drawn from the comparative analysis of the data on the structure of high-olefin copolymers are as follows.

The simultaneous polymerization of olefins with complex catalysts yields real copolymers rather than the mixtures of homopolymers. The monomer sequences in the copolymers have the same stereoregular structure as the corresponding homopolymers. Hence, both homopolymerization and copolymerization of olefins take place on the same stereospecific sites.

The distribution of monomer units in the copolymers in the majority of cases is in satisfactory agreement with the $r_1 r_2$ values calculated from $f-F$ correlations. Therefore the possible distribution of active sites in respect to r_1 and r_2 values does not significantly alter the distribution of monomer units.

The degree of randomness of the olefin copolymers varies significantly, depending on the combination of olefins and the catalytic system chosen. The most striking feature is that some olefin copolymers have an evident strong tendency toward block formation. Such copolymers were obtained mainly with highly specific catalytic systems.

The examination of $r_1 r_2$ values for hundreds of different comonomers polymerized by different mechanisms (2) reveals that in the overwhelming majority of cases these $r_1 r_2$ values are close to or less than 1; very few examples of ionic processes were found with $r_1 r_2 > 1$ (169). For this reason the appearance of a significant number of cases with $r_1 r_2 > 1$ can be regarded as characteristic of complex catalysis. The mentioned tendency is especially pronounced when the comonomers have alkyl groups of different size (ethylene–4-methylpentene-1, propylene–butene-1, propylene–styrene, propylene–4-methylpentene-1, propylene–vinylcyclohexane). On the other hand, when the alkyl groups are of similar bulkiness (4-methylpentene-1–vinylcyclohexane, 4-methylpentene-1–3-methylbutene-1, vinylcyclohexane–styrene), the copolymers obtained are mainly random or have a tendency to alternation.

Special mechanisms have been proposed (47, 118, 145, 169, 170), attributing this phenomenon to the peculiarities of the growth of the helix chain on the surface of the heterogeneous catalyst.

VIII. Acknowledgements

I express my sincere appreciation to my colleagues of the Institute of Chemical Physics, the Institute of Petrochemical Synthesis, Moscow Petrochemical Plant and PLASTPOLYMER (Leningrad) for the preparation and characterization of different copolymers which have been studied in our laboratory over many years and were discussed here. I am also pleased to thank Prof. C. Tosi (MONTECATINI-EDISON) for the helpful discussion of various aspects of IR spectroscopy of copolymers.

The author is indebted to Mary Lekich for help in translating the paper.

IX. References

1. Mayo, F. R., Walling, C.: Chem. Rev. **46**, 191 (1950).
2. Ham, G. E. (Ed.): Copolymerisation. Interscience 1964.
3. — J. Polymer Sci. **45**, 169, 177 (1960); **61**, 9 (1962).
4. Miller, R. L.: J. Polymer Sci. **46**, 303 (1960); **57**, 975 (1962).
5. Ham, G. E.: J. Macromol. Sci. A **5**, 453 (1971).
6. Pyun Chong, Wha: J. Polymer Sci. A **2**, **8**, 1111 (1970).
7. Klesper, E.: J. Polymer Sci. A **1**, **8**, 1191 (1970).
8. Harwood, J. H.: Am. Chem. Soc. Polymer Preprints **8**, 199 (1967).
9. Kissin, Yu. V., Visen, E. I.: Vysokomol. Soyedin. A **16**, 1385 (1974).
10. — Tsvetkova, V. I., Chirkov, N. M.: Vysokomol. Soyedin. A **9**, 1374 (1967).
11. — — — Europ. Polymer J. **8**, 529 (1972).
12. Allegra, G.: Makromol. Chem. **117**, 24 (1968).
13. Natta, G., Dall'Asta, G., Mazzanti, G., Ciampelli, F.: Kolloid-Z. **182**, 50 (1962).
14. Cosse, P., Arlman, E. J.: J. Catalysis **3**, 80, 99 (1964).
15. Rodriguez, L., Van Loy, N. M.: J. Polymer Sci. A **1**, **4**, 1905 (1966).
16. Kissin, Yu. V., Chirkov, N. M.: Europ. Polymer J. **6**, 525 (1970).
17. — Tsvetkova, V. I., Chirkov, N. M.: Vysokomol. Soyedin. A **9**, 1104 (1967).
18. Shelden, R. A., Fueno, T., Tsunetsugu, T., Furukawa, J.: J. Polymer Sci. B **3**, 23 (1965).
19. Tosi, C.: Advan. Polymer Sci. **5**, 451 (1968).
20. Askerov, K. M., Seidov, N. M., Abdullaev, R. D.: Vysokomol. Soyedin. A **15**, 932 (1973).
21. Tosi, C., Valvassori, A., Ciampelli, F.: Europ. Polymer J. **4**, 107 (1968).
22. Kinsinger, J. B., Fisher, T., Wilson, N. J.: J. Polymer Sci. B **5**, 285 (1967).
23. Kissin, Yu. V., Mezhikovskiy, S. M., Chirkov, N. M.: Europ. Polymer J. **6**, 267 (1970).
24. Brück, D., Hummel, D. O.: Makromol. Chem. **163**, 281 (1973).
25. Markevitch, A.: Vysokomol. Soyedin. A **9**, 502 (1967).
26. Firsov, A. N., Meshkova, I. N., Kostrova, N. D., Chirkov, N. M.: Vysokomol. Soyedin. **8**, 1860 (1966).
27. Tosi, C.: Makromol. Chem. **170**, 231 (1973).
28. Fineman, M., Ross, S. D.: J. Polymer Sci. **5**, 269 (1950).
29. Joshi, R. M., Joshi, S. G.: J. Macromol. Sci. A **5**, 1329 (1971).
30. Tidwell, P. V., Mortimer, G. A.: J. Polymer Sci. A **3**, 369 (1965); J. Macromol. Sci. C **4**, 281 (1970).
31. Braun, D., Brendlein, W., Mott, G.: Europ. Polymer J. **9**, 1007 (1973).
32. Tosi, C., Ciampelli, F.: Advan. Polymer Sci. **12**, 87 (1973).
33. Wegemer, N. J.: J. Appl. Polymer Sci. **14**, 573 (1970).
34. Tosi, C., Lachi, M. P., Pinto, A.: Makromol. Chem. **120**, 225 (1968).
35. Seidov, N. M., Dalin, M. A.: Vysokomol. Soyedin. B **13**, 80 (1971).
36. Natta, G., Mazzanti, G., Valvassori, A., Pajaro, G.: Chim. Ind. (Milano) **41**, 764 (1959).
37. Slonaker, D. F., Comb, R. L., Coover, H. W.: J. Macromol. Sci. A **1**, 539 (1967).
38. Hayashi, I., Ohno, K.: Chem. High Pol. (Japan) **22**, 446 (1965).
39. Sakagushi, F., Tsuji, W., Kitamaru, R.: Chem. High Pol. (Japan) **24**, 493 (1967).
40. Huff, T., Bushman, C. J., Cavender, J. V.: J. Appl. Polymer Sci. **8**, 825 (1964).

41. Gehrke, K., Bledzki, A., Ulbricht, J.: Plaste u. Kautschuk **17**, 251 (1970).
42. Turner-Jones, A.: J. Polymer Sci. B**3**, 591 (1965).
43. Uiliem Kho, Kissin, Yu. V., Goldfarb, Yu. Ya., Krentsel, B. A.: Vysokomol. Soyedin. A**14**, 2229 (1972).
44. Kho, Uiliem, Kissin, Yu. V., Kleiner, V. I., Krentsel, B. A., Stotskaya, L. L., Zakharyan, R. Z.: Europ. Polymer J. **9**, 315 (1973).
45. Kissin, Yu. V., Goldfarb, Yu. Ya., Krentsel, B. A., Kho, Uiliem: Europ. Polymer J. **8**, 487 (1972).
46. — Yakobson, F. I., Amerik, V. V., Krentsel, B. A.: Vysokomol. Soyedin. A**14**, 3 (1972).
47. Dankovicŝ, A., Kissin, Yu. V.: Vysokomol. Soyedin. A**12**, 802 (1970).
48. Bakayutov, N. G., Kissin, Yu. V. et al.: Vysokomol. Soyedin (in press).
49. Goldenberg, A. L., Zapletnyak, V. M., Ilchenko, P. A.: Polymer Spectroscopy, p. 195. Kiev: Naukova Dumka 1968.
50. Carlini, C., Ciardelli, F., Pino, P.: Makromol. Chem. **119**, 244 (1968).
51. Khodjaeva, V. L., Mamedova, V. A.: Vysokomol. Soyedin. A**9**, 447 (1967).
52. Entelis, S. G., Vainshtein, E. F., Kumpanenko, I. V.: Vysokomol. Soyedin. A**11**, 2615 (1969).
53. Kissin, Yu. V.: Thesis, Moscow, 1965.
54. Kumpanenko, I. V., Kazanski, K. S.: Advances in polymer chemistry and physics. Khimiya, Moscow, 1973, p. 64.
55. Zerbi, G.: Pure Appl. Chem. **26**, 499 (1971).
56. Tosi, C., Zerbi, G.: Chim. Ind. (Milano) **55**, 334 (1973).
57. Kumpanenko, I. V.: Thesis, Moscow, 1971.
58. Zbinden, R.: Infrared spectroscopy of high polymers. New York-London: Academic Press 1964.
59. Uno, T., Machida, K.: Spectrochim. Acta A**24**, 1741 (1968).
60. Kumpanenko, I. V., Kazanski, K. S.: Vysokomol. Soyedin. A**15**, 594 (1973).
61. Laceŝ, J.: J. Molec. Spectroscopy **30**, 167 (1969).
62. Zerbi, G., Ciampelli, F., Zamboni, V.: J. Polymer Sci. C**7**, 141 (1963).
63. Kobayashi, M., Tsumura, K., Tadokoro, H.: J. Polymer Sci. A**2**, **6**, 1493 (1968).
64. Folt, V. L., Shipman, J. J., Krimm, S.: J. Polymer Sci. **61**, 17 (1962).
65. Kobayashi, M., Akita, K., Tadokoro, H.: Makromol. Chem. **118**, 324 (1968).
66. Shteinbak, V. Sh., Kissin, Yu. V., Yakobson, F. I., Mezhikovski, S. M., Amerik, V. V.: 18 USSR Polymer Conf., Kazan, 1973, p. 49.
67. Krimm, S.: Adv. Polymer Sci. **2**, 51 (1960).
68. Natta, G., Valvassori, A., Ciampelli, F., Mazzanti, G.: J. Polymer Sci. A**3**, 1 (1965).
69. Bucci, G., Siminazzi, T.: J. Polymer Sci. C**7**, 203 (1964).
70. Ciampelli, F., Tosi, C.: Spectrochim. Acta **24**A, 2157 (1968).
71. Druchel, H. V., Ellerbe, J. S., Cox, R. C., Lane, L. H.: Analyt. Chem. **40**, 370 (1968).
72. Kissin, Yu. V., Tsvetkova, V. I., Chirkov, N. M.: Vysokomol. Soyedin. A**10**, 1092 (1968).
73. Miyamoto, T., Inagaki, H.: J. Polymer Sci. A**2**, **7**, 963 (1969).
74. Magrupov, M. A., Gafurov, I.: Uzbekh. Khim. Zh. **4**, 36 (1968).
75. Peraldo, M., Cambini, M.: Spectrochim. Acta **21**, 1509 (1965).
76. — Natta, G., Zambelli, A.: Makromol. Chem. **89**, 273 (1965).
77. Zerbi, G., Gussoni, M., Ciampelli, F.: Spectrochim. Acta **23**A, 301 (1967).
78. Zambelli, A., Natta, G., Pasquon, I.: J. Polymer Sci. C**4**, 411 (1963).
79. Luongo, J. P., Salovey, R.: J. Polymer Sci. A**2**, **4**, 997 (1966); B**3**, 513 (1965).

80. Goldbach, G., Peitscher, G.: J. Polymer Sci. B**6**, 783 (1968).
81. Natta, G.: Makromol. Chem. **35**, 94 (1960).
82. Nishioka, T., Yanagisawa, K.: Chem. High Pol. (Japan) **19**, 667, 671 (1962).
83. Tadokoro, H., Kitazawa, K., Nizakura, S., Murahashi, S.: Bull. Chem. Soc. Japan **34**, 1209 (1961).
84. Ukita, M.: Bull. Chem. Soc. Japan **39**, 742 (1966).
85. Rubin, I. D.: J. Polymer Sci. A**2**, **5**, 1323 (1968).
86. Berdnikova, M. P., Kissin, Yu. V., Chirkiv, N. M.: Vysokomol. Soyedin. **5**, 63 (1963).
87. Morero, D., Mantica, E., Ciampelli, F., Sianesi, D.: Nuovo Cimento **15**, 122 (1960).
88. Liang, C. Y., Krimm, S.: J. Polymer Sci. **27**, 241 (1958).
89. Tadokoro, H., Nishiyama, N., Nazakura, S., Murahashi, S.: J. Polymer Sci. **36**, 553 (1959).
90. Onishi, T., Krimm, S.: J. Appl. Phys. **32**, 2320 (1961).
91. Braun, D., Betz, N., Kern, W.: Naturwissenschaften **46**, 444 (1959).
92. Takeda, M., Iimura, K., Yamada, A., Imamura, Y.: Bull. Chem. Soc. Japan **32**, 1150 (1959).
93. Natta, G., Corradini, P., Cesari, M.: R. C. Accad. Lincei **22**, 11 (1957).
94. Helms, J. B., Ghalla, G.: J. Polymer Sci. A**2**, **10**, 761 (1972).
95. Johnson, L. F., Heatley, F., Bovey, F.: Macromolecules **3**, 175 (1970).
96. Helms, J. B., Ghalla, G.: J. Polymer Sci. A**2**, **10**, 1147 (1972).
97. Lim, D., Kolinsky, M., Petranek, J., Doskočhil, D., Schneider, B.: J. Polymer Sci. B**4**, 645 (1966).
98. Yanagisawa, K., Ashikari, N., Kanemitsu, T., Nishioka, A.: Chem. High Pol. (Japan) **21**, 312, 319 (1964).
99. Luongo, J. P.: J. Polymer Sci. B**5**, 281 (1967).
100. Yamada, S., Kanakahara, Y., Kitahara, S.: Chem. High Pol. (Japan) **23**, 521 (1966).
101. Bovey, F. A.: High resolution NMR of macromolecules. New York-London: Academic Press 1972.
102. Shelden, R. A.: J. Polymer Sci. A**2**, **7**, 1111 (1969).
103. Tsuruta, T.: J. Polymer Sci. **1972**, 179.
104. Ito, K., Yamashita, Y.: J. Polymer Sci. A**3**, 2165 (1965); B**3**, 625 (1965).
105. Flory, P. J.: Principles of polymer chemistry, p. 570. Ithaca, N.Y.: Cornell University Press 1953.
106. Flory, P. J.: Trans. Faraday Soc. **51**, 848 (1955).
107. Richardson, M. J., Flory, P. J., Jackson, J. B.: Polymer **4**, 221 (1963).
108. Schaefgen, J. P.: J. Polymer Sci. **38**, 549 (1959).
109. Yakobson, F. I., Amerik, V. V., Ivanyukov, D. V., Petrova, V. F., Kissin, Yu. V., Krentsel, B. A.: Vysokomol. Soyedin. A**13**, 2699 (1971).
110. Baker, C. H., Mandelkern, L.: Polymer **7**, 7 (1966).
111. Mandelkern, L.: Chem. Rev. **56**, 903 (1956).
112. Allegra, G., Bassi, I. W.: Advan. Polymer Sci. **6**, 549 (1969).
113. Reding, F. P., Walter, F. R.: J. Polymer Sci. **37**, 555 (1959).
114. Turner-Jones, A.: Polymer **6**, 249 (1965).
115. Natta, G.: Makromol. Chem. **35**, 94 (1960).
116. Belov, G. P., Belova, V. A., Raspopov, L. N., Kissin, Yu. V., Brikenshtein, K. A., Chirkov, N. M.: Polymer J. **3**, 681 (1972).
117. van Schooten, J., Duck, E. W., Berkenbosch, R.: Polymer **2**, 357 (1961).
118. Coover, H. W.: J. Polymer Sci. C**4**, 1511 (1963).
119. Kinner, W., Schmoltke, R.: Plaste u. Kautschuk **15**, 807 (1968).

120. Tosi, C., Valvassori, A., Ciampelli, F.: Europ. Polymer J. **5**, 575 (1969).
121. Natta, G., Valvassori, A., Mazzanti, G.: Chim. Ind. (Milano) **40**, 896 (1958).
122. Takegami, Y., Suzuki, T., Kondo, T., Mitani, K.: Chem. High Pol. (Japan) **29**, 199 (1972).
123. Zambelli, A., Tosi, C., Sacchi, C.: Macromolecules **5**, 649 (1972).
124. Goldenberg, A. L.: Thesis, Leningrad, 1969.
125. Natta, G., Mazzanti, G., Valvassori, A., Sartori, G., Morero, D.: Chim. Ind. (Milano) **42**, 125 (1960).
126. Zambelli, A., Lèty, A., Tosi, C., Pasquon, I.: Makromol. Chem. **115**, 73 (1968).
127. Gaylord, N. G., Mark, H. F.: Linear and stereoregular addition polymers. New York: Interscience 1959.
128. Buchick, R. I.: J. Polymer Sci. A **3**, 2047 (1965).
129. Cosewith, C., VerStrate, G.: Macromolecules **4**, 482 (1971).
130. Zerbi, G., Gussoni, G., Ciampelli, F.: Spectrochim. Acta **24** A, 301 (1967).
131. Ciampelli, F., Valvassori, A.: J. Polymer Sci. C **16**, 377 (1967).
132. Takegami, Y., Suzuki, T.: J. Polymer Sci. B **9**, 109 (1971).
133. Crain, W. O., Zambelli, A., Roberts, J. D.: Macromolecules **4**, 330 (1971).
134. Zambelli, A., Gatti, G., Sacchi, C., Crain, W. O., Roberts, J. D.: Macromolecules **4**, 475 (1971).
135. Carman, C. J., Wilkes, C. E.: Rubber Chem. Technol. **44**, 781 (1971).
136. VerStrate, G., Wilchinsky, Z. W.: J. Polymer Sci. A **2**, **9**, 127 (2971).
137. Michajlov, L., Zugenmeier, P., Cantow, H.-J.: Polymer **9**, 325 (1968).
138. Davison, S., Taylor, G. L.: Brit. Polymer J. **4**, 65 (1972).
139. Holdsworth, R. J., Keller, A.: J. Polymer Sci. B **5**, 605 (1967).
140. Bodily, D., Wunderlich, B.: J. Polymer Sci. A **2**, **4**, 25 (1966).
141. Jackson, J. F.: J. Polymer Sci. A **1**, 2119 (1963).
142. Michajlov, L., Cantow, H.-J., Zugenmaier, P.: Polymer **12**, 70 (1971).
143. Shnell, G.: Ber. Bunsen-Ges. Phys. Chem. **70**, No. 3 (1966).
144. Matkovski, P. E., Belov, G. P. *et al.*: Vysokomol. Soyedin. A **12**, 2286 (1970).
145. Turner-Jones, A.: Polymer **7**, 23 (1966).
146. Tokuzumi, T.: Chem. High Pol. (Japan) **25**, 721 (1968).
147. Seidov, N. M., Dalin, M. A., Kyazimov, S. M.: Dokl. Acad. Nauk USSR **164**, 826 (1965).
148. Efendieva, T. Z., Koptev, D. A.: Vysokomol. Soyedin. A **10**, 227 (1968).
149. Seidov, N. M., Koptev, D. A., Agakisheva, M. Y.: Preprints of International Symposium on Macromolecules, Helsinki, Vol. 2, Section 1, 797 (1972); Azerbaijan Khem. Zh., No. 5 (1967).
150. — Dalin, M. A.: Dokl. Acad. Nauk USSR **170**, 396 (1966).
151. — — Preprints of Internstional Symposium on Macromolecular Chemistry, Budapest II, 349 (1969).
152. Agakisheva, M. Y., Seidov, N. M., Kuliev, T. A.: Azerbaijan. Khim. Zh. 111 (1969).
153. Seidov, N. M., Kyazimov, S. M., Guseinova, D. F.: International Symposium on Macromolecules, Helsinki, Vol. 2, Section 1, 803 (1972).
154. Kobayashi, S., Watanabe, H., Nishioka, A.: Chem. High Pol. (Japan) **23**, 103 (1966).
155. Tadokoro, H., Kobayashi, M., Ukita, M., Yasuzuki, K., Murahashi, S.: J. Chem. Phys. **42**, 1432 (1965).
156. Coover, H. W., McDonnell, R. L., Joyner, F. B., Slonaker, D. F., Combs, R. L.: J. Polymer Sci. A **1**, **4**, 2563 (1966).
157. Kawai, W.: Bull. Chem. Soc. Japan **38**, 879 (1965).

158. Zambelli, A., Natta, G., Pasquon, I.: J. Polymer Sci. C **4**, 411 (1963).
159. Ashikari, N., Kanemitsu, T., Yanagisawa, K., Nakagawa, K., Okamoto, H., Kobayashi, S., Nishioka, A.: J. Polymer Sci. A **2**, 3009 (1964).
160. Kobayashi, S., Kato, Y., Watanabe, H., Nishioka, A.: J. Polymer Sci. A **1**, 4, 245 (1966).
161. Razuvaev, G. A., Minsker, K. S., Shapiro, I. Z.: Vysokomol. Soyedin. **4**, 1833 (1962).
162. Chernovskaya, R. P., Lebedev, V. P., Minsker, K. S., Razuvaev, G. A.: Vysokomol. Soyedin. **6**, 1313 (1964).
163. Seidov, N. M., Guseinov, T. I., Abasov, A. Z., Kasimov, K. G., Yurjeva, G. N., Mamedova, Y. M.: Vysokomol. Soyedin. B **14**, 641 (1972).
164. Hayashi, I., Ichikawa, R.: J. Chem. Soc. Japan, Ind. Chem. Sec. **70**, 728, 733, 735 (1967).
165. Otsu, T., Nagahama, H., Endo, K.: J. Polymer Sci. B **10**, 601 (1972).
166. — Shimizu, A., Itakura, K., Imoto, H.: Makromol. Chem. **123**, 289 (1969).
167. Atarashi, Yu.: J. Chem. Soc. Japan, Ind. Chem. Sec. **68**, 2487 (1965).
168. Anderson, I. H., Burnett, G. M., Tait, P. J. T.: Proc. Chem. Soc. 225 (1960).
169. — — — J. Polymer Sci. **56**, 391 (1962).
170. — — Geddes, W. C.: Europ. Polymer J. **3**, 161, 171, 181 (1967).
171. Schafden, J. R.: J. Polymer Sci. **38**, 549 (1959).
172. Natta, G., Allegra, G., Bassi, I. W., Carlini, C., Chienlini, E., Montagnoli, G.: Macromolecules **2**, 311 (1969).
173. Isaacson, R. B., Kirshenbaum, I., Klein, I.: J. Appl. Polymer Sci. **9**, 933 (1965).
174. Yanagisawa, K.: Chem. High Pol. (Japan) **22**, 58 (1965).
175. Natta, G., Corradini, P., Sianesi, D., Morero, D.: J. Polymer Sci. **51**, 527 (1961).
176. — Danusso, F., Sianesi, D.: Makromol. Chem. **30**, 238 (1959).
177. Ciardelli, F., Carlini, C., Montagnoli, G.: Macromolecules **2**, 296 (1969).
178. — Benedetti, E., Montagnoli, G., Lucarini, L., Pino, P.: Chem. Commun. 285 (1965).
179. Tsvetkova, V. I., Chirkov, N. M. et al.: Problemy kinetiki i kataliza (USSR) **12**, 253 (1968).
180. Vizen, E. I., Kissin, Yu. V.: Vysokomol. Soyedin. A **11**, 1774 (1969).
181. Tosi, C.: Chim. Ind. (Milano) **53**, 459 (1971).
182. Meskova, I. N., Dubnikova, I. L., Vizen, E. I., Chirkov, N. M.: Vysokomol. Soyedin. B **11**, 486 (1969).
183. Edgecombe, F. H. C.: Nature **198**, N 4885, 1085 (1963).
184. Natta, G., Mazzanti, G., Valvassori, A., Sartori, G., Barbagello, A.: J. Polymer Sci. **51**, 429 (1961).
185. Lipman, R. D. A.: Am. Chem. Soc. Polymer Preprints **8**, 369 (1967).
186. Laputte, R., Guyot, A.: Makromol. Chem. **129**, 234 (1969).
187. Hajashi, A.: J. Chem. Soc. Japan, Ind. Chem. Sec. **66**, 1356 (1963).
188. Ketley, A. D.: J. Polymer Sci. B **1**, 121 (1963).
189. Leitan, O.: Thesis, Moscow, 1969.
190. Shteinbak, V. Sh., Amerik, V. V., Yakobson, F. I., Ivanyukov, D. V., Krentsel, B. A.: Vysokomol. Soyedin. A **15**, 1621 (1973).
191. Jung, K. A., Schnecko, H.: Makromol. Chem. **154**, 227 (1972).
192. Schnecko, H., Dost, W., Kern, W.: Makromol. Chem. **121**, 159 (1969).
193. Overberger, C. G., Mujarnichi, K.: J. Polymer Sci. A **1**, 2021 (1963).
194. Gehrke, K., Bledski, A., Schmidt, B., Ulbricht, J.: Plaste u. Kautschuk **18**, 87 (1971).
195. Bledski, A., Müller, B., Ulbricht, J.: Plaste u. Kautschuk **20**, 593 (1973).

Received March 20, 1974

Die Makromolekulare Chemie

An International Journal
of Macromolecular Chemistry and Physics

Founded by Hermann Staudinger, Nobel Prize Laureate in Chemistry
Editor: Werner Kern

Die Makromolekulare Chemie

- carries papers from authors working at all major universities and research centers throughout the world;
- publishes only original papers that meet high standards of excellence and significance;
- is clearly divided into: "Chemistry of Macromolecules", "Physical Chemistry of Macromolecules", and "Physics of Macromolecules";
- provides for speedier publication of important news as "Short Communications";
- gives "Summeries" in English as well as in the original language (65% of the papers being written in English, 25% in German and 10% in French).

Die Makromolekulare Chemie is published in 12 issues a year totalling about 3600 pages. Rate for 1974 (volume 175): sfr. 1260.— / DM 1056.—.
Back issues and complete-back sets are available. Please inquire.

Hüthig & Wepf Verlag, Eisengasse 5, CH-4001 Basel

Springer-Verlag
Berlin Heidelberg New York
München Johannesburg London
Madrid New Delhi
Paris Rio de Janeiro Sydney Tokyo Utrecht Wien

POLYMER CHEMISTRY

By Professor B. Vollmert,
Polymer Institute, University of Karlsruhe, Germany
Translated from the German by E. H. Immergut, New York
With 630 figures. XVII, 652 pages. 1973
Cloth DM 72.—; US $29.40 ISBN 3-540-05631-9
Prices are subject to change without notice

This book gives a comprehensive coverage of the synthesis of polymers and their reactions, structure, and properties. The treatment of the reactions used in the preparation of macromolecules and in their transformation into cross-linked materials is particularly detailed and complete. The book also gives an up-to-date presentation of other important topics, such as enzymatic and protein synthesis, solution properties of macromolecules, polymer in the solid state. The content and presentation of Professor Vollmert's book is more encompassing than most existing treatises, and its numerous figures and tables convey a wealth of data, never, however, at the expense of intellectual clarity or educational value.
The presentation is mainly on a fundamental and general level and yet the reader—student or professional—is gradually and almost casually introduced to all important natural and synthetic polymers. Complicated phenomena are explained with the aid of the simplest available examples and models in order to ensure complete understanding. However, the reader is also encouraged to think for himself and even to criticize the author's point of view. All of the chapters have been revised and enlarged from the German edition, and many of the sections are entirely new.

Contents: Introduction. — Structural Principles. — Synthesis and Reactions of Macromolecular Compounds. — The Properties of the Individual Macromolecule. — States of Macromolecular Aggregation.